T0144489

RIVER PUBLISHERS SERIES IN MATHEMATICAL AND ENGINEERING SCIENCES

Series Editors

MANGEY RAM
Graphic Era University
India

ALIAKBAR MONTAZER HAGHIGHI
Prairie View Texas A&M University
USA

TADASHI DOHI
Hiroshima University
Japan

Indexing: All books published in this series are submitted to the Web of Science Book Citation Index (BkCI), to SCOPUS, to CrossRef and to Google Scholar for evaluation and indexing.

Mathematics is the basis of all disciplines in science and engineering. Especially applied mathematics has become complementary to every branch of engineering sciences. The purpose of this book series is to present novel results in emerging research topics on engineering sciences, as well as to summarize existing research. It engrosses mathematicians, statisticians, scientists and engineers in a comprehensive range of research fields with different objectives and skills, such as differential equations, finite element method, algorithms, discrete mathematics, numerical simulation, machine leaning, probability and statistics, fuzzy theory, etc.

Books published in the series include professional research monographs, edited volumes, conference proceedings, handbooks and textbooks, which provide new insights for researchers, specialists in industry, and graduate students.

Topics covered in the series include, but are not limited to:

- Advanced mechatronics and robotics
- Artificial intelligence
- Automotive systems
- Discrete mathematics and computation
- Fault diagnosis and fault tolerance
- Finite element methods
- Fuzzy and possibility theory
- Industrial automation, process control and networked control systems
- Intelligent control systems
- Neural computing and machine learning
- Operations research and management science
- Optimization and algorithms
- Queueing systems
- Reliability, maintenance and safety for complex systems
- Resilience
- Stochastic modelling and statistical inference
- Supply chain management
- System engineering, control and monitoring
- Tele robotics, human computer interaction, human-robot interaction

For a list of other books in this series, visit www.riverpublishers.com

AN INTRODUCTION TO TENSOR ANALYSIS

Dr. Bipin Singh Koranga

M.Sc., Ph.D. (IITB), MIPA,

Department of Physics,

Kirori Mal College, Delhi

Dr. Sanjay Kumar Padaliya

M.Sc., Ph.D. (K.U),

Department of Mathematics,

SGRR (PG) College, Dehradun, India

River Publishers

Published, sold and distributed by:

River Publishers

Alsbjergvej 10

9260 Gistrup

Denmark

www.riverpublishers.com

ISBN: 978-87-7022-581-6 (Hardback)

    978-87-7022-580-9 (Ebook)

©2020 River Publishers

Preface

I feel great pleasure in bringing out the first edition of the book "An Introduction to Tensor Analysis". This book has been written especially in accordance with the latest and modified syllabus framed for under-graduate and postgraduate students. A reasonably wide coverage in sufficient depth has been attempted. The book contains sufficient number of problems. We hope that if a student goes through all these, he or she would appreciate and enjoy the subject. The work is dedicated to our past students whose inspiration motivated us to do this work without any stress and strain. The authors wish to acknowledge his indebtedness to numerous authors of those books, which were consulates during the preparation of the matter. We feel great pleasure to express deepest sense of gratitude, respect and honor to Prof. Uma Sankar, IIT Bombay and Prof. Mohan Narayan, Mumbai University for conferring valuable guidance and their patronly behavior. We would like to thank several of our colleagues in Kirori Mal College and SGRR(PG) College, Dehradun. We would like to thank Dr. Vinod Kumar, University of Lucknow, Dr. Imran Khan, Ramjas College, Dr. Shushil Kumar, Hindu College, bearing for making useful suggestions during the preparation of manuscript.

Errors might have crept in here and there inspite of the care to avoid them. We will be very grateful for bringing any errors to our notice. Suggestions or criticisms towards the further improvement of the book shall be gratefully acknowledged. We shall appreciate receiving comments and suggestions, which can be sent to the below emails: bipiniitb@rediffmail.com, spadaliya12@rediffmail.com

Thanks to Prof. Angelo Galanty, University of L'Aquila and Dr. Rakesh Joshi, Australia's Global University for encouragement. We are deeply grateful to our family members, who have always been a source of inspiration for us. In the last, but not the least, we are thankful to the publisher of this book.

Dr. Bipin Singh Koranga

Dr. Sanjay Kumar Padaliya

Syllabus

Cartesian Tensors-Transformation of Coordinates. Einstein's Summation Convention. Relation between Direction Cosines. Tensors. Algebra of Tensors. Sum, Differences and Product of Two Tensors. Contraction. Quotient Law of Tensors. Symmetric and Anti-symmetric Tensors. Pseudo tensors. Invariant Tensors: Kronecker and Alternating Tensors. Association of Anti-symmetric Tensor of Order two and Vectors. Vector Algebra and Calculus using Cartesian Tensors: Scalar and Vector Products, Scalar and Vector Triple Products. Differentiation. Gradient, Divergence and Curl of Tensor Fields. Vector Identities. Tensorial Formulation of Analytical Solid Geometry: Equation of a Line. Angle Between Lines. Projection of a Line on another Line. Condition for Two Lines to be Coplanar. Foot of the Perpendicular from a Point on a Line. Rotation Tensor (No Derivation). Isotropic Tensors. Tensorial Character of Physical Quantities. Moment of Inertia Tensor. Stress and Strain Tensors : Symmetric Nature. Elasticity Tensor. Generalized Hooke's Law

General Tensors-Transformation of Coordinates. Contravariant and Covariant Vectors. Contravariant, Covariant and Mixed Tensors. Kronecker Delta and Permutation Tensors. Algebra of Tensors. Sum,

Difference and Product of Two Tensors. Contraction. Quotient Law of Tensors. Symmetric and Anti-symmetric Tensors. Metric Tensor. Reciprocal Tensors. Associated Tensors. Christoffel Symbols of First and Second Kind and their Transformation Laws. Covariant Derivative. Tensor Form of Gradient, Divergence and Curl.

Contents

1

Introduction

The concept of tensor has its origin in the development of differential geometry by Gauss, Riemann and Christoffel. The investigation of relation, which remains valid, when we change from one coordinate system to another coordinate system is the main aim of Tensor Analysis. The law of physics can not be dependent on the frame of reference, that the physicist chooses for the purpose of explanining Tensor Analysis as the mathematical background in which such a law can be formulated. In particular Einstein found it as excellent tool for the presentation of the general theory of relativity. Its application to most branches of Theoretical Physics.

Any natural law giving as it does a relation between different physical entities is mathematically formulated as relation between the set of number representation the entities natural law is that. It does not depend upon the reference frame chosen for the reperesenation of the entities and the mathmatical formulation of the law should desirably be such that it remains in the same form when the reference frame is changed in any manner.

Tensor formulation is very compact and a good deal of clarity is achieved. This subject had been originally formulated by G. Ricci which come out as a very natural tool for the description of his general theory of relativity. It has also since found useful for studies in differential geometry, mechanics, electromagnitism and elasticity. The orthogonal rectilinear co-ordinate system only relevent in relation to the definition of Cartesian Tensor. General Tensor which are related to the consideration of general curvilinear system of coordinates. It will be seen that every general tensor is a cartesian tensor but the converse is not true. Tensor analysis deals with entities and properties of the choice of reference frames. Tensor is an ideal tool for the study of natural law.

1.1 Symbols Multi-Suffix

A set of the numbers are denoted by a one suffix symbol such as a_i, b_i and c_i, where i takes as its value the three numbers 1, 2, 3. Again a set of nine numbers considered as a trial of triads is denoted by a two-suffix symbol such as

$$a_{ij}, b_{ij} \text{ and } c_{ij} \text{ etci} = 1, 2, 3; \text{ j} = 1, 2, 3.$$

The set of numbers given by any two suffix symbol a_{ij} can be exhibited in the form of a square array or a matrix.

$$\begin{pmatrix} a_{11} & a_{12} & a_{13} \\ a_{21} & a_{22} & a_{23} \\ a_{31} & a_{32} & a_{33} \end{pmatrix}$$

where i and j respectively denote row and column suffixes.

A suffix may not take up three and instead may assume any number of values of n. This will be the case, when we consider of n dimensions. Thus, in the n dimension Euclidean space, the one-suffix symbol a_i will denote a set of n numbers.

$$a_i....a_{3......}a_n$$

and the two-suffix symbol a_{ij} will denote a set of n^2 numbers, which may be arranged in the form of a square array as follows

$$\begin{pmatrix} a_{11} & a_{12} & a_{13} \\ a_{21} & a_{22} & a_{23} \\ a_{31} & a_{32} & a_{33} \end{pmatrix}$$

The number of suffixes in a multi-suffix symbol is called the order and the number of values assumed by each suffix is the dimension of the set.

1.2 Summation Convention

We have to consider the sum of numbers, which constitute certain sub-set of the sets of numbers given by multi-suffix symbols, when these sub-sets arise on giving all possible equal values to some two of the suffixes. Thus, we may have the sums.

$$\sum_{i=1}^{i=n} a_{ij} = a_{11} + a_{22} + a_{33} + a_{nn}$$

$$\sum_{i=1}^{i=n} a_{iji} = a_{1j1} + a_{2j2} + a_{3j3} + a_{njn}$$

in the relation to the sets a_{ij} and a_{ijk}, respectively. It is found convenient to drop the sign $\sum$ of summation and regard the presence of two equal suffixes itself as denoting summation. Thus, we shall write,

$$a_{ij} = a_{11} + a_{22} + a_{33} + a_{nn}$$

$$a_{iji} = a_{1j1} + a_{2j2} + a_{3j3} + a_{njn}$$

$$a_{ijij} = a_{1j1j} + a_{2j2j} + a_{3j3}j + a_{njnj}$$

It will be seen that a_{ij} denotes a single number and a_{iji} are one-suffix and two-suffix symbols respectively. Thus, a symbol having two identical suffixes stand for the sum obtained by giving all possible values to the identical suffixes and without altering the remaining suffixes. The subject of tensor deals with the problem of the formulation of relations between varous entites in form which remain unchanged, when we pass from the on system of coordinate to another. As the facts of invariance of an equation is essentially related to the possible types of system of co-ordinate which reference to which the equation remains invariant, we must necessary define the various system of coordinate system to be taken into account relative to which the invariance is to be occured. We now state that the our basic purpose of this book is the study of the invariance of equation relative to the rectangular coordinate system in the three-dimension Euclidean system. The book can be considered to be in two-parts- cartesian tensor and general tensor.

References

[1] Harold Jeeffreys (1931), Cartesian Tensors, PP(1-16), Cambridge University Press(New York)

[2] David C. Kay, Theory and Problem of Tensors Calculus, PP(1-3), McGraw Hill, Washinton, D.C.

[3] Shanti Narayan (1961), Cartesian Tensors, PP(1-12), S chand, New Delhi.

[4] DE Bourine and PC Kendell (1967), Vector Analysis and Cartesian Tensor, PP(245-257), Chapman &Hall.

[5] Barry Spain (1960), Tensor Calculus, PP(1-55), Dover Publication, Newyork.

[6] A.J. McConnell (1960), Application of Tensor Analysis, PP(1-9), Khosla Publication, New Delhi.

[7] Zefer Ahson (2000), Tensor Analysis with Applications, Anamaya Publisher, New Delhi.

[8] U.C. De (2008), Tensor Calculus, PP(1-9), Narosa Publishing House, New Delhi.

[9] J.D Anand (2013), Mathmatical Physics V, PP(151-161), Hai Anand Publication, New Delhi.

[10] J.K Goyal (1998), Tensor Calculus and Riemanninan Geometry, Pragati Prakashan, Merrut.

[11] Charlie Harper, (1970), Introduction to Mathmatical Physics, PP(255-275), Prentice Hall India Learning Private Limited, Delhi.

[12] M.L Boas (1966), Mathmatical Methods in the Physical Sciences, PP(496-526), Wiley.

[13] A.W Joshi, (1975), Matrices and Tensor In Physics, PP(187-219), New Age International, Publishers, New Delhi.

2

Cartesian Tensor

2.1 Introduction

The subject of tensor deals with the problem of the formulation of relation between various entities in form which remain invariant, when we pass from one system of coordinate to another. Invariant of equation is necessary related to the possible types of system of coordinates with reference to which the equation remains invariant. We now state that the primary purpose of this chapter is the study of the invariant of equation relative to the total of rectangular coordinate system in the three-dimension Euclidean space. We have to start with the consideration of manner in which the sets representing various entities are transformation when we pass from one system of rectangular coordinates to another. A tensor may be a physical entity that can be described as a tensor only with respect to some manner of its representation by means of multi-suffix sets associated with different systems of axes such that the sets associated with different systems of coordinate obey the transformation law for tensor. We have employed

7

suffix notation for tensors of any order, we could also employ single letters such as A, B to denote tensors.

2.2 Transformation of Coordinates

If we have two systems of rectangular coordinate axes OX, OY, OZ; OX', OY', OZ' having the same origin O such that the direction of cosines of the lines OX', OY', OZ' relative to the system OXYZ are

$$l_1, m_1, n_1; l_2, m_2, n_2; l_3, m_3, n_3.$$

The following transformation equations express x', y', z' in terms of x, y, z and vice versa:

$$x' = l_1 x + m_1 y + n_1 z,$$

$$y' = l_2 x + m_2 y + n_2 z,$$

$$z' = l_3 x + m_3 y + n_3 z,$$

$$x = l_1 x' + l_2 y' + l_3 z',$$

$$y = m_1 x' + m_2 y' + m_3 z',$$

$$z = n_1 x' + n_2 y' + n_3 z',$$

where x', y', z' and x, y, z are the coordinates of the same point relative to the two systems of coordinate axes. Now we write these equations in a compact form in terms of multiple suffix sets and summation convention. We say that $x_i, \bar{x}_i$ are the coordinates of a point P relative to the two systems of axes, the range of each suffix being 1,2,3. Let l_{ij} denote the cosine of the angle between OX_I *and* $O\bar{X}_J$. The direction cosines of $O\bar{X}_1, O\bar{X}_2, O\bar{X}_3$ relative to the system $OX_1X_2X_3$ are given by the columns and those of OX_1, OX_2, OX_3 relative to $O\bar{X}_1\bar{X}_2\ X_3$ are given by the row of the square matrix

$$\begin{bmatrix} l_{11} & l_{12} & l_{13} \\ l_{21} & l_{22} & l_{23} \\ l_{31} & l_{32} & l_{33} \end{bmatrix},$$

The equation of coordinate transformation can, be written in terms of suffix notation as follows;

$$\bar{x}_1 = l_{11}x_1 + l_{21}x_2 + l_{31}x_3, \qquad x_1 = l_{11}\bar{x}_1 + l_{12}\bar{x}_2 + l_{13}\bar{x}_3,$$

$$\bar{x}_2 = l_{12}x_1 + l_{22}x_2 + l_{32}x_3, \qquad x_1 = l_{21}\bar{x}_1 + l_{22}\bar{x}_2 + l_{23}\bar{x}_3,$$

$$\bar{x}_1 = l_{13}x_1 + l_{23}x_2 + l_{33}x_3, \qquad x_1 = l_{31}\bar{x}_1 + l_{32}\bar{x}_2 + l_{33}\bar{x}_3,$$

The summation convention equations are written as;

$$\bar{x}_1 = l_{i1}x_i, \qquad x_1 = l_{1j}\bar{x}_j,$$

$$\bar{x}_2 = l_{i2}x_i, \qquad x_2 = l_{2j}\bar{x}_j,$$

$$\bar{x}_3 = l_{i3}x_i, \qquad x_2 = l_{3j}\bar{x}_j,$$

We can re-write these compactly as a single equations in the form

$$\bar{x}_j = l_{i1}x_i, \qquad\qquad x_i = l_{ij}\bar{x}_j,$$

which are complete equivalents of the equation of coordinate transformation from one system to the another.

2.3 Relations Between the Direction Cosines of Three Mutually Perpendicular Straight Lines

The direction cosines of the three mutually perpendicular straight lines $O\bar{X}_1, O\bar{X}_2, O\bar{X}_3$ relative to the system $OX_1X_2X_3$ are

$$l_{11}, l_{21}, l_{31}; l_{12}, l_{22}, l_{32}; l_{13}, l_{23}, l_{33},$$

They are connected by the following relations

$$l_{11}l_{11} + l_{21}l_{21} + l_{31}l_{31} = 1, \qquad l_{11}l_{12} + l_{21}l_{22} + l_{31}l_{32} = 0,$$

$$l_{12}l_{12} + l_{21}l_{22} + l_{32}l_{32} = 1, \qquad l_{11}l_{13} + l_{22}l_{23} + l_{32}l_{33} = 0,$$

$$l_{13}l_{13} + l_{23}l_{23} + l_{33}l_{33} = 1, \qquad l_{13}l_{11} + l_{23}l_{21} + l_{33}l_{31} = 0,$$

In summation convention, we can write these relations as

$$l_{i1}l_{i1} = 1, \qquad\qquad l_{i1}l_{i2} = 0,$$

$$l_{i2}l_{i2} = 1, \qquad\qquad l_{i2}l_{i3} = 0,$$

$$l_{i3}l_{i3} = 1, \qquad\qquad l_{i3}l_{i1} = 0.$$

Finally all above equations can write as single equation

$$l_{ij}l_{ik} = \delta_{jk},$$

where δ_{jk} is the Kroneocker delta.

2.4 Transformation of Velocity Components

It is known that any given velocity can be represented by means of its three components along three mutually perpendicular lines and the three components characterize the velocity components. Clearly the component changes as we pass from one system of mutually perpendicular lines to another. We consider two rectangular systems

$$OX_1X_2X_3, \quad O\bar{X}_1\bar{X}_2\bar{X}_3,$$

$l_i, \bar{l}_j$ are the direction cosines of the line of action of the velocity and v denotes the magnitude of the velocity. If v_i, v_i denote the components of the velocity relative to the two systems of axes, we have

$$v_i = vl_i, \bar{v}_j = v\bar{l}_j. \tag{2.1}$$

If now draw a line through the origin O parallel to the line of action of the velocity, then the coordinate of the point P on this line at unit distance from O relative to the two systems of axes are,

$$l_i, l_j.$$

The equation of coordinate transformation, we have

$$\bar{l}_j = l_{ij}l_i, \quad l_i = l_{ij}\bar{l}_j. \tag{2.2}$$

From Eq.(2.1) and Eq.(2.2), we have

$$\bar{v}_j = l_{ij}v_i, \quad v_i = l_{ij}\bar{v}_j. \tag{2.3}$$

Thus, we see that the equation (2.3) for the transformation of velocity.

2.5 First-Order Tensors

Any entity represented by a set of three numbers (component) relative to a system of rectangular axes is called a first-order tensor, if its components a_i, $\bar{a}_j$ relative to any two systems of rectangular axes $OX_1X_2X_3$, $O\bar{X}_1\bar{X}_2\bar{X}_3$ are connected by the relation

$$\bar{a}_j = l_{ij}a_i; \tag{2.4}$$

with l_{ij} being the cosine of the angle between $O\bar{X}_I$ and $O\bar{X}_J$.

It can be seen that the components of first-order tensor obey the same transformation law as the coordinate of a point. A tensor of first order is also called a Vector. Any entity represented by a single number such that the same number represents the entity irrespective of any underlying system of axes is called a tensor of zero order. A tensor of order zero is also called a scalar.

2.6 Second-Order Tensors

Considering any two tensors of first order and letting

$$a_i, b_j; \qquad \bar{a}_p \bar{b}_q,$$

be the components of the same relative to two different systems of axes, we have

$$\bar{a}_p = l_{ip} a_i,$$

$$\bar{b}_q = l_{jp} b_j,$$

where l_{ip}, l_{jq} have usual meanings. We have

$$\bar{a}_p \bar{b}_q = l_{ip} a_i l_{jq} b_j$$

$$= l_{ip} l_{jq} a_i b_j. \tag{2.5}$$

The right-hand side of Eq.(2.5) denotes the sum 9 term, obtained by giving the dummy suffixes i, j all possible pairs of value. Any entity by two-suffix set relatively to a system of rectangular axes is called a second order, if the sets a_{ij}, $\bar{a}_{pq}$ representing the entity relative to any two systems of rectangular axes

$OX_1X_2X_3$, $O\bar{X}_1\bar{X}_2\bar{X}_3$ are connected by the relation

$$\bar{a}_{pq} = l_{ip}l_{jq}a_{ij},$$

Any entity consider by a set with, m suffixes relative to a system of rectangular coordinate axes is called a tensor of order m, if the sets a_{ijkl}............,$\bar{a}_{pqrs}$.......... represented by the entity relative to any two systems of rectangular axes $OX_1X_2X_3$, $O\bar{X}_1\bar{X}_2\bar{X}_3$ are connected by the relation

$$\bar{a}_{pqrs}............. = l_{ip}l_{jq}l_{rs}l_{ls}.......a_{ijkl}...............,$$

we say that a_{ijkl}...... are the components of the tensor relatively to the rectangular system of axes $OX_1X_2X_3$.

2.7 Notation for Tensors

In elementery geometrical treatment, where a vector is defined as a directed line segment. It is usual to denote a vector by a single faced letter, such as a and b. In the present analytical method approach

2.8 Algebraic Operations on Tensors

We now define operation on tensor such that by the means of different operations, we can construct new tensors.

2.8.1 Sum and Difference of Tensors

If $a_{ijkl......}$, $b_{ijkl......}$ are two tensors of the same order, then

$$c_{ijkl......} = a_{ijkl.....} + b_{ijkl..........}$$

is also a tensor of the same order.

Let

$$a_{ijkl......}, \; b_{ijkl......} \text{ and } \bar{a}_{pqrs......}, \; \bar{b}_{pqrs......}$$

be the components of the given tensors relative to two systems $OX_1X_2X_3, \;\; O\bar{X}_1\bar{X}_2\bar{X}_3$. We write

$$c_{ijkl......} = a_{ijkl.....} + b_{ijkl..........}$$

$$\bar{c}_{pqrs......} = \bar{a}_{pqrs.....} + \bar{b}_{pqrs..........}$$

Let, I_{ij} denote the cosine of the angle between OX_i and $O\bar{X}_j$. The statement proved now showing that

$$\bar{c}_{pqrs......} = I_{ip}I_{jq}I_{kr}I_{ls}a_{ijkl.....} \tag{2.6}$$

As $a_{ijkl......}$, $b_{ijkl......}$ are tensors, we have

$$\bar{a}_{pqrs......} = I_{ip}I_{jq}I_{kr}I_{ls}a_{ijkl.....} \tag{2.7}$$

$$\bar{b}_{pqrs......} = I_{ip}I_{jq}I_{kr}I_{ls}b_{ijkl.....} \tag{2.8}$$

Adding 2.6 and 2.7, we obtained 2.8. Hence the new tensor is said to be the sum of the given tensors. The case of the diffference of tensors can be obtain in a similar way.

2.8.2 Product of Tensors

If $a_{ijkl......}, b_{ijkl......}$ are two tensors of the orders α and β, respectively, then

$$c_{ijkl......pqrs......} = a_{ijkl......}b_{pqrs......}$$

is a tensor of order $\alpha + \beta$.

Let

$$a_{ijkl......}, b_{pqrs......}, \bar{a}_{i_1 j_1 k_1 l_1......}, \bar{b}_{p_1 q_1 r_1 s_1}$$

be the components of the tensor relative to the two systems $OX_1X_2X_3, O\bar{X}_1O\bar{X}_2\bar{X}_3$.

We write

$$c_{ijkl.....pqrs.....} = a_{ijkl.....}b_{pqrs......}$$

$$\bar{c}_{i_1 j_1 k_1 l_1.....p_1 q_1 r_1 s_1..........} = \bar{a}_{i_1 j_1 k_1 l_1}\bar{b}_{p_1 q_1 r_1 s_1..........}$$

The new tensor obtained is called the product of the tensor. The product of two tensor is a tensors whose order is the sum of the orders of the given tensors.

2.9 Quotient Law of Tensors.

This law is useful in establishing tensorial character of given entities. If there be an entity representable by a multi-suffix set a_{ij}, relative to any given system of rectangular axes and if a_{ij}, b_i is a vector whatsoever, then a_{ij} is a tensor of order two.

We write

$$a_{ij}b_i = c_j \qquad (2.9)$$

so that c_j is a vector.

Let

$$a_{ij}, b_i, c_j$$

and

$$\bar{a}_{pq}\bar{b}_p, \bar{c}_q$$

be components of the given entity and the two vectors relative to two of axes $OX_1X_2X_3$, $O\bar{X}_1\bar{X}_2\bar{X}_3$

As given, we have

$$a_{ij}b_i = c_j \ldots\ldots\ldots \qquad (2.10)$$

$$a_{pq}\bar{b}_p = c_q \ldots\ldots\ldots \qquad (2.11)$$

Also b_i, c_j being vectors, we have

$$c_q = \bar{l}_{jq} c_j \quad ---- \quad (2.12)$$

$$b_i = l_{iq} \bar{b}_p \quad ------ \quad (2.13)$$

From these equation, we obtain

$$\bar{a}_{pq} \bar{b}_p = \bar{c}_q$$

$$= l_{jq} c_j$$

$$= l_{jq} a_{ij} b_i$$

$$= l_{jq} a_{ij} l_{ip} \bar{b}_p$$

$$= l_{ip} l_{jq} a_{ij} \bar{b}_p$$

$$(\bar{a}_{pq} - l_{ip} l_{jq} a_{ij}) \bar{b}_p = 0. \quad (2.14)$$

As the vector $\bar{b}_p$ is arbitrary, we consider three vectors, whose components relative to $O\bar{X}_1 \bar{X}_2 \bar{X}_3$.

$$1, 0, 0; 0, 1, 0; 0, 0, 1.$$

For these vectors, we have from (2.14),

$$\bar{a}_{1q} - l_{i1} l_{jq} a_{ij} = 0, \bar{a}_{2q} - l_{i2} l_{jq} a_{ij} = 0, \bar{a}_{3q} - l_{i3} l_{jq} a_{ij} = 0,$$

These are equivalent to

$$\bar{a}_{pq} - l_{ip}l_{jq}a_{ij} = 0,$$

$$\bar{a}_{pq} = l_{ip}l_{jq}a_{ij},$$

So that the components of the given entity obey the tensorial transformation law.

2.10 Contraction Theorem

If

$$a_{ijk\ldots\ldots\ldots}$$

is a tensor of order m, then the set obtained on identifying any two suffixes is a tensor of order (m-2).

Let

$$\bar{a}_{ijkl}\ldots\ldots\ldots\ldots\bar{a}_{pqrs}\ldots\ldots\ldots\ldots,$$

be the components of the given tensors relative to two systems

$$OX_1X_2X_3, \ O\bar{X}_1\bar{X}_2\bar{X}_3$$

so that, we have

$$\bar{a}_{pqrs}\ldots\ldots\ldots\ldots = l_{ip}l_{jp}l_{rs}l_{ls}\ldots\ldots\ldots\ldots a_{ijkl}. \qquad (2.15)$$

Let us consider the second and fourth suffixes and we then write

$$a_{ijkl}.................... = c_{ik}.............., \overline{a}_{pqrs}......................... = \overline{c}_{pr}.$$

The theorem now amounts to show that

$$\overline{c}_{pr}............. = l_{ip}l_{kr}........................c_{ik}.$$

Identifying q with s in Eq.(2.15), we get

$$\overline{a}_{pqrs}..................... = l_{ip}l_{js}l_{kr}l_{ls}...............a_{ijkl}. \qquad (2.16)$$

Now we have

$$i_{js}i_{ls} = \delta_{jl} \qquad (2.17)$$

and

$$\delta_{jl}a_{ijkl.... } = a_{ijkl.....}. \qquad (2.18)$$

From (2.16),(2.17) and (2.18), we have

$$\overline{a}_{pqrs} = l_{ip}l_{kr}......a_{ijkl},$$

$$\overline{c}_{pr}............. = l_{ip}l_{kr}........................c_{ik}.$$

Hence the theorem.

2.11 Symmetric and Skew-Symmetric Tensor

Let

$$a_{ijkl......}, \overline{a}_{pqrs......}$$

be the components of a tensor relative to two systems $OX_1X_2X_3$, $O\bar{X}_1\bar{X}_2\bar{X}_3$. We shall prove that if $a_{ijkl.....}$ is symmetric (skew-symmetric) in any two suffixes, then so is also $a_{pqrs.....}$ in the same two suffixes.

We have

$$\overline{a}_{pqrs........} = l_{ip}l_{jq}l_{kr}l_{ls........}a_{ijkl}. \tag{2.19}$$

Now suppose that

$$a_{ijkl........}$$

is a symmetric tensor in the second and fourth suffixes. Interchanging q and s on the two-side of (2.19). We obtain

$$\overline{a}_{pqrs} = l_{ip}l_{js}l_{kr}l_{ls........}a_{ijkl.........} \tag{2.20}$$

As j and l are dummies, we can interchange them. Then interchanging j and l on the right of (2.19). We obtained

$$\overline{a}_{pqrs} = l_{ip}l_{js}l_{kr}l_{ls........}a_{ijkl.........},$$

$$= l_{ip}l_{jq}l_{kr}l_{ls........}a_{ijkl.........}, \tag{2.21}$$

with the set $a_{ijkl....}$ being symmetric in the second and fourth suffixes from (2.19) and (2.21), we have

$$\overline{a}_{pqrs........} = \overline{a}_{psrq.......}$$

A tensor is said to be symmetric (skew-symmetric) in any two suffixes, if its Components relative to every coordinate system are symmetric (skew-symmetric) in the two suffixes in question. For example, we may see that if u_i, v_j be any two vectors, then the two second-order tensors.

$$u_i v_j + u_j v_i, \qquad u_i v_j - u_j v_i,$$

are respectively, symmetric and skew-symmetric. If u_i, v_j and w_k are any three vectors then the three second-order tensors

$$u_i v_j w_k + u_j v_k w_i + u_k v_i w_j + u_i v_k w_j + u_j v_i w_k + u_k v_j w_i$$

and

$$u_i v_j w_k + u_j v_k w_i + u_k v_i w_j - u_i v_k w_j - u_j v_i w_k - u_k v_j w_i$$

are respectively symmetric and skew-symmetric.

We shall now define two special tensor, alternate tensor and kronecker tensor, of order-three and order-two respectively.

2.12 Alternate Tensor

Consider an abstract entity of order 3 and dimension 3 such that its components relative to every system of coordinate axes are the same and given by ϵ_{ijkl}, where, any two of i, j, k are equal

$\epsilon_{ijkl} = 1$, $\;if\;$ i, j, k is a cyclic permutation of 1,2,3

$\epsilon_{ijkl} = -1$, $\;if\;$ i, j, k is a anti-cyclic permutation of 1,2,3.

Thus, for unequal values of the suffixes, we have

$$\epsilon_{123} = \epsilon_{231} = \epsilon_{312} = 1, \epsilon_{132} = \epsilon_{213} = \epsilon_{321} = -1.$$

It will be shown that the entity is a tensor of order three. Let $OX_1X_2X_3$, $O\bar{X}_1\bar{X}_2\bar{X}_3$ be two Systems of rectangular axes. Consider, now the expression

$$l_{ip}l_{iq}l_{kr}\epsilon_{ijk}. \tag{2.22}$$

For any given system of values of p, q, r, s, the expression 1 consists of a sum of $3^3 = 27$ terms of which only 6 are non-zero for the other 21 terms corresponding to case when at least two of i, j, k are equal. It may now be seen that the expression 1 is the same as the determinant.

$$\begin{pmatrix} i_{1p} & i_{2p} & i_{3p} \\ i_{1q} & i_{2q} & i_{3q} \\ i_{1r} & i_{2r} & i_{3r} \end{pmatrix}.$$

From the elementary properties of determinants, we see that this determinant

$D = 0,$ if any two of p, q, r are equal.

$D = 1,$ if p, q, r is a cyclic permutation of 1, 2, 3.

$D = -1,$ if p, q, r is a anti-cyclic permutation of 1, 2, 3.

Thus, we see that the components of the given entity in any two-systems of rectangular coordinate axes satisfy the tensorial transformation equation so that the entity is a tensor. This tensor is known as the Alternate tensor. Clearly, alternate tensor is a skew-symmetric tensor. The alternate tensor will always be denoted by the symbol ϵ_{ijk}.

2.13 Kronecker Tensor

Consider an entity of order two such that its components relativel to every coordinate system of axes are the same and given by δ_{ij}, where

$$\delta_{ij} = 0, if\ i \neq j$$

$$\delta_{ij} = 1, if\ i = j.$$

It will now be shown that this entity is a tensor. In the usual notation, consider the expression

$$l_{ip}l_{jq}\delta_{ij},$$

This is equal to

$$l_{1p}l_{1q} + l_{2p}l_{2q} + l_{3p}l_{3q} = 0, if\ p \neq q$$

$$l_{1p}l_{1q} + l_{2p}l_{2q} + l_{3p}l_{3q} = 1, if\ p = q.$$

Thus, we see that the given tensor is Kronecker Tensor. The tensorial character of δ_{ij} can also be seen to follow easily from the quotient law of tensors as follows. We have, for any arbitrary vector a_i,

$$\delta_{ij}a_i = a_i,$$

so that for any arbitray vector a_i, $\delta_{ij}a_i$ is also a Vector. Hence, by Quotient law δ_{ij} is a tensor of order two. A tensor has the same set of components relative to every system of coordinate axes is called an Isotropic tensor. Alternate tensor and kronecker tensor are both isotropic tensors of orders three and two respectively.

2.14 Relation Between Alternate and Kronecker Tensors

We shall now prove the following important relation between the alternate tensor and kronecker tensor;

$$\epsilon_{ijm}\epsilon_{klm} = \delta_{ik}\delta_{jl} - \delta_{il}\delta_{jk.}$$

Here each side is a tensor of order 4 so that the tensor equalities requried to be proved is equivalent to a set of 81 scalar equalities,
We have to prove that

$$\epsilon_{ij1}\epsilon_{kl1} + \epsilon_{ij2}\epsilon_{kl2} + \epsilon_{ij3}\epsilon_{kl3} = \delta_{ik}\delta_{jl} - \delta_{il}\delta_{jk}.$$

We may easily verify that when i, j are equal or when k, l are equal.

In the case of unequal values of i, j and unequal values of k, l, we may again easily verify that if the pair of unequal values of i, k, j is different from the pair of unequal values of k, l, Then

$$\epsilon_{ijm}\epsilon_{klm} = \delta_{ik}\delta_{jl} - \delta_{il}\delta_{jk}.$$

Thus, we are left to consider the possibilities when i, j and k, l take the pairs of values

$$1, 2; 1, 3; 2, 3, 2; 1, 3, 1; 3, 2.$$

Consider the first case so that we may have

$$i = 1, j = 2, k = 1, l = 2; i = 1, j = 2, k = 2, l = 1$$

$$i = 2, j = 1, k = 1, l = 2; i = 2, j = 1, k = 2, l = 1.$$

In the next case, we have

$$\epsilon_{ijm}\epsilon_{klm} = 1, \delta_{ik}\delta_{jl} - \delta_{il}\delta_{jk} = 1; \epsilon_{ijm}\epsilon_{klm} = -1, \delta_{ik}\delta_{jl} - \delta_{il}\delta_{jk} = -1$$

$$\epsilon_{ijm}\epsilon_{klm} = -1, \delta_{ik}\delta_{jl} - \delta_{il}\delta_{jk} = -1; \epsilon_{ijm}\epsilon_{klm} = 1, \delta_{ik}\delta_{jl} - \delta_{il}\delta_{jk} = 1.$$

The result may no be easily seen to be true in the other case.

2.15 Matrices and Tensors of First and Second Orders

We shall now make some observations and some facts to show how the manipulation with tesnor of first and second orders can be brought into relationship with the algebra of matrices. Consider first any vector. Its components a_i relative to any system of axes may be written in the form of a row or a column matrix as

$$\left(\begin{matrix} a_1 & a_2 & a_3 \end{matrix} \right) \qquad or \qquad \left(\begin{matrix} a_1 \\ a_2 \\ a_3 \end{matrix} \right).$$

We shall be writing

$$[a_i] = \left[\begin{matrix} a_1 & a_2 & a_3 \end{matrix} \right] \qquad or \qquad [a_i] = \left[\begin{matrix} a_1 \\ a_2 \\ a_3 \end{matrix} \right].$$

Consider now any second-order tensor. Its components a_{ij} relative to any system of axes can be written in the form of a matrix such that a_{ij} occurs at the intersection of the ith row and the jth column; the first elements to the left denoting row and the second denoting column. Thus, we shall write

$$[a_{ij}] = \left[\begin{matrix} a_{11} & a_{12} & a_{13} \\ a_{21} & a_{22} & a_{23} \\ a_{31} & a_{32} & a_{33} \end{matrix} \right].$$

However, it should be clearly understood that different matrices may correspond to the same tensor depending on the system of coordinate axes to which the tensor. A matrix obtained by interchanging the row and columns of a given matrix is called the transpose of the same. The transpose of $[a_{ij}]$ will be denoted by

$[a_{ij}]'$. Sum of two matrices of the same type is the matrix whose elements are the sums of the corresponding elements of the two matrices.

2.16 Product of Two Matrices

If $[a_{ij}]$ and $[b_{ij}]$ are two $m \times n$ and $n \times p$ matrices, then their product is the $m \times p$ matrix $[c_{ik}]$, where

$$c_{ik} = a_{i1}b_{1k} + a_{i2}b_{2k} + a_{i3}b_{3k} + \ldots\ldots\ldots\ldots a_{in}b_{nk}$$

$$= a_{ij}b_{jk}$$

Considering the summation convention.

We shall now consider the different points. Consider a vector whose components relative to the systems $OX_1X_2X_3$ and $O\bar{X}_1\bar{X}_2\bar{X}_3$ are a_i and a_j. Then, l_{ij} having the usual meaning, we have

$$\bar{a}_j = l_{ij}a_i = a_i l_{ij}.$$

In the matrix form, this equality is equivalent to

$$
\begin{bmatrix} \bar{a_1} & \bar{a_2} & \bar{a_3} \end{bmatrix} = \begin{bmatrix} a_1 & a_2 & a_3 \end{bmatrix} \begin{bmatrix} l_{11} & l_{12} & l_{13} \\ l_{21} & l_{22} & l_{23} \\ l_{31} & l_{32} & l_{33} \end{bmatrix}
$$

or

$$[\bar{a_j}] = [\bar{a_i}] \, [l_{ij}].$$

Consider now a tensor of order 2 whose components relative to $OX_1X_2X_3$ and $O\bar{X}_1\bar{X}_2\bar{X}_3$ are a_{ij} and a_{pq}. Then, we have

$$\bar{a}_{pq} = l_{ip}l_{jq}a_{ij}$$

In matrix form this is equivalent to

$$\bar{a}_{pq} = \left[l'_{pi}\right][a_{ij}][l_{jq}]$$

where

$$l'_{pi} = l_{ip}$$

$$
\begin{bmatrix} \bar{a}_{11} & \bar{a}_{12} & \bar{a}_{13} \\ \bar{a}_{21} & \bar{a}_{22} & \bar{a}_{23} \\ \bar{a}_{31} & \bar{a}_{32} & \bar{a}_{33} \end{bmatrix}
$$

$$
= \begin{bmatrix} l_{11} & l_{21} & l_{13} \\ l_{21} & l_{22} & l_{23} \\ l_{31} & l_{23} & l_{33} \end{bmatrix} \begin{bmatrix} a_{11} & a_{12} & a_{13} \\ a_{21} & a_{22} & a_{23} \\ a_{31} & a_{32} & a_{33} \end{bmatrix} \begin{bmatrix} l_{11} & l_{12} & l_{13} \\ l_{21} & l_{22} & l_{23} \\ l_{31} & l_{32} & l_{33} \end{bmatrix}.
$$

Consider now two tensors a_{ij}, b_{pq}. The matrix $[a_{ij}b_{jq}]$ of this inner product is given by

$$[a_{ij}b_{jq}] = [a_{ij}][b_{jq}].$$

This matrices of other inner products of the two given tensors can also be written down similarly. Consider for example the inner product

$$a_{ij}b_{kj}$$

We write $b'_{jk} = b_{kj}$. Thus,

$$[a_{ij}b_{kj}] = \left[a_{ij}b'_{jk}\right] = [a_{ij}]\left[b'_{jk}\right]$$

$$= \begin{bmatrix} a_{11} & a_{12} & a_{13} \\ a_{21} & a_{22} & a_{23} \\ a_{31} & a_{32} & a_{33} \end{bmatrix} \begin{bmatrix} b_{11} & b_{12} & b_{13} \\ b_{21} & b_{22} & b_{23} \\ a_{31} & a_{32} & a_{33} \end{bmatrix}.$$

Consider now two vectors u_i, v_j. We have

$$[u_i, v_j] = \begin{bmatrix} u_1 \\ u_2 \\ u_3 \end{bmatrix} \begin{bmatrix} v_1 & v_2 & v_3 \end{bmatrix} = \begin{bmatrix} u_1v_1 & u_1v_2 & u_1v_3 \\ u_2v_1 & u_2v_2 & u_2v_3 \\ u_3v_1 & u_3v_2 & u_3v_3 \end{bmatrix}$$

$$[u_iv_j] = \begin{bmatrix} u_1 & u_2 & u_3 \end{bmatrix} \begin{bmatrix} v_1 \\ v_2 \\ v_3 \end{bmatrix} = \begin{bmatrix} u_1v_1+ & u_2v_2+ & u_3v_3 \end{bmatrix}.$$

It is usual to note that

$$[\delta_{ij}] = \begin{bmatrix} 1 & 0 & 0 \\ 0 & 1 & 0 \\ 0 & 0 & 1 \end{bmatrix}.$$

2.17 Scalar and Vector Inner Product

2.17.1 Two Vectors

We should remember that associated with any tensor or system of tensors, we may have other tensors, which arise through the operations of multiplication and contraction as performed on the given tensors and other known tensors, especially isotropic tensors. Infacts the tensor notation supplies a very direct method for setting up invariant which are entities independent of coordinate system, and tensor are such entities. We shall now define scalar and vector products of two vectors which are infact tensors of zeros-order and order-one associated with the given vector in the manner referred to above. It will be seen the the two types of products arises in tensor notation. In this connection, we observe that in seeting from two vectors u_i, v_j. We are here interested in setting up other tensors which are either scalars or vectors, tensors of order zero or one.

2.17.2 Scalar Product

The scalar $u_i v_i$ is called the scalar product of the two vectors u_i, v_i. Thus, the scalar product

$$u_i v_j = u_1 v_1 + u_2 v_2 + u_3 v_3. \tag{2.23}$$

2.17.3 Vector Product

The vector $\epsilon_{ijkl} u_i v_j$ is called the product of the vectors u_i, v_j taken in this order. It may be easily seen that the components of this vector product are

$$u_2 v_3 - u_3 v_1, u_3 v_1 - u_1 v_3, u_1 v_2 - u_2 v_1. \qquad (2.24)$$

2.18 Tensor Fields

A tensor field or a tensor point function is said to be defined when there is given a law which is associated to each point region of space tensor of the same order. Thus a tensor field $a_{ij......}$ of any order is defined if the components $a_{ij.....}$ are a function of x_1, x_2, x_3.

2.18.1 Gradient of Tensor Field

Let u be a scalar point function so that there is a value associated with each point of a given region of space. Thus, if $OX_1 X_2 X_3$ and $O\bar{X}_1 \bar{X}_2 \bar{X}_3$ be any two system, then we can look upon u indifferently as a function of x_1, x_2, x_3 and of $\bar{x}_1, \bar{x}_2, \bar{x}_3$ or we can say $\bar{x}_1$ and $\bar{x}_p$ which are the coordinates of any point P relative to the system of axes. For any point P $\bar{x}_1$ and $\bar{x}_p$ are different but the values of u_i are the same.

Considering now the sets of first-order $\frac{\partial u}{\partial x_1}, \frac{\partial u}{\partial x_2}$, we have

$$\frac{\partial u}{\partial \bar{x}_p} = \frac{\partial u \partial x_1}{\partial x_1 \partial \bar{x}_p} + \frac{\partial u \partial x_2}{\partial x_2 \partial \bar{x}_p} + \frac{\partial u \partial x_3}{\partial x_3 \partial \bar{x}_p}$$

$$= \frac{\partial u \partial x_i}{\partial x_i \partial \bar{x}_p}. \qquad (2.25)$$

Also, we have

$$\overline{x}_p = l_{ip}x_i,$$

or

$$x_i = l_{ip}\overline{x}_p.$$

By differentiation, we have

$$\frac{\partial x_i}{\partial \overline{x}_p} = l_{ip}. \tag{2.26}$$

From Eq.(2.25) and Eq.(2.23), we obtained

$$\frac{\partial u}{\partial \overline{x}_p} = l_{ip}\frac{\partial u}{\partial x_i}.$$

We write

$$\frac{\partial u}{\partial x_i} = a_i, \quad \frac{\partial u}{\partial \overline{x}_p} = \overline{a}.$$

We have

$$\overline{a}_p = l_{ip}a_i.$$

Thus we see that a_i, $\frac{\partial u}{\partial x_i}$ is a tensor of order one a vector. Components $\frac{\partial u}{\partial x_i}$ and $\frac{\partial u}{\partial \overline{x}_p}$ relative to two systems of axes $OX_1X_2X_3$, $O\overline{X}_1\overline{X}_2\overline{X}_3$ obey the tensorial transformation law. This vector is called the gradient of the scalar u.

2.18.2 Divergence of Vector Point Function

The scalar of the gradient of a vector point function is called the divergence of the point function. Thus, if u_i is a vector function so that

$$u_{ij} = \frac{\partial u_i}{\partial x_j},$$

is its gradient, then

$$u_{ij} = \frac{\partial u_i}{\partial x_j} = \frac{\partial u_1}{\partial x_1} + \frac{\partial u_2}{\partial x_2} + \frac{\partial u_3}{\partial x_3}.$$

is called the divergence of u_i denoted by the symbol $Div u_i$. The divergence of any tensor is defined as gradient of the contracted on the first and the last indices. Thus

$$Div.u_{ijkl} = u_{ijkl}................_i.$$

2.18.3 Curl of Vector Point Function

The vector of the gradient of vector point function is called the curl of the point function. Thus, if u_i is a vector function so that

$$u_{ij} = \frac{\partial u_i}{\partial x_j},$$

is its gradient, then the vector of the tensor, the vector

$$\epsilon_{ijkl} u_{ij}$$

is called the curl or rotor of u_i denoted by the symbol curl u_i. Clearly, the components of curl u_i are

$$\frac{\partial u_3}{\partial x_2} - \frac{\partial u_2}{\partial x_3}, \frac{\partial u_1}{\partial x_3} - \frac{\partial u_3}{\partial x_1}, \frac{\partial u_2}{\partial x_1} - \frac{\partial u_1}{\partial x_2}.$$

The Skew-symmetric second-order tensor associated with the vector curl u_i is

$$u_{ij} - u_{ji}.$$

2.19 Tensorial Formulation of Gauss's Theorem

If F is a continuously differentiable vector point function and S is a closed surface enclosing a region V, then

$$S \int F_i n_i dS = V \int F_{i,i} dV, \tag{2.27}$$

where n is the unit outward normal vector.

2.20 Tensorial Formulation of Stoke's Theorem

If F is a continuously differentiable vector point function and S is a closed surface enclosing by a curve C, then

$$C \int F_i t_i ds = S \int \epsilon_{ijkl} F_{k,j} n_i ds, \tag{2.28}$$

where the unit vector n at any point S is drawn in the sense in which a right handed screw would rotated in the sense of description of C.

2.21 Exercise

Prove that

(1) $\delta_{ij} = 3$.

(2) $\delta_{ij}\epsilon_{ijk} = 0$.

(3) $\epsilon_{ijk}\epsilon_{ijk} = 6$.

(4) $\epsilon_{pjk}\epsilon_{kpj} = 2\delta_{pq}$.

(5) Show that

$$
\epsilon_{ijk}\epsilon_{kpj} = \begin{pmatrix}
\delta_{il} & \delta_{im} & \delta_{in} \\
\delta_{jl} & \delta_{jm} & \delta_{jn} \\
\delta_{kl} & \delta_{km} & \delta_{kn}
\end{pmatrix}
$$

and deduce the relation between alternate tensor and Kronecker tensor.

(6) If l_{ij} are the direction cosines of a transformation of axes prove that

$$
\epsilon_{ikm}l_{mn} = \epsilon_{jln}l_{ij}l_{kl},
$$

$$
l_{ij} = \frac{1}{2}\epsilon_{ikm}\epsilon_{jln}l_{kl}l_{mn}.
$$

(7) if a_{ij} is a skew-symmetric second-order tensor, prove that

$$
(\delta_{ij}\delta_{ik} + \delta_{il}\delta_{jk})a_{ik} = 0
$$

(8) if a_{ijk} is a tensor, prove that a_{jki} is also a tensor.

References

[1] Harold Jeeffreys (1931), Cartesian Tensor, PP(1-66), Combridge University Press (New York).

[2] David C. Kay, Theory and Problem of Tensor Calculus, PP(1-3), McGraw Hill, Washinton, D.C.

[3] Shanti Narayan (1961), Cartesian Tensor, PP(37-51), S. Chand, New Delhi.

[4] Barry Spain (1960), Tensor Calculus, PP(1-55), Dover Publication, Newyork.

[5] Zefer Ahson(2000). Tensor Analysis with Application, Anamaya Publisher, New Delhi.

3

Tensor in Physics

3.1 Kinematics of Single Particle

Velocity, momentum and acceleration; consider any particle of mass m and let at time t, the particle be at point P, where coordinates relative to a system of axes OXYZ are x_i, then the three components of the velocity $\frac{dx_i}{dt}$ are said to be the components of the velocity of the moving particle at time t relative to the system OXYZ. Let now $OX'Y'Z'$ be another system of axes fixed relative to OXYZ so that with usual notation, we have

$$\overline{x_j} = l_{ij}x_i, \tag{3.1}$$

where l_{ij} are constants, differentating Eq.(3.1), with respect to time we have

$$\frac{dx'_i}{dt} = l_{ij}\frac{dx_i}{dt}.$$

This transformation law, for the components $\frac{dx'_i}{dt}$ and $\frac{dx_i}{dt}$. Show that the velocity of a particle is a vector.

3.1.1 Momentum

The mass m of a particle being scalar

$$m\frac{dx_i}{dt},$$

called momentum is also a vector.

3.1.2 Acceleration

The three components of acceleration

$$\frac{dx_i^2}{dt^2}, \quad i = 1 - 3,$$

Now component of transformation equation

$$\frac{d^2x_i}{dt^2} = l_{ij}\frac{dx_i^2}{dt^2},$$

where, $\frac{dx_i^2}{dt^2}$ and $\frac{d^2x_i}{dt^2}$ satisfied the transformation law for tensor.

3.1.3 Force

Experimentally it was found that one of the reference frames for which the Newtons's second law of motion holds to a good degree of approximation is the reference frame attached to the so called "Fixed Stars". In any case, we shall say that any reference frame for which the law is true is called an inertial frame and we shall assume that the underlying reference frame is always an inertial frame.

If X_i is the components of the force along the axes acting on a particle of mass m, where position at time t is x_i, then by Newton's second law of motion with appropriate choice of units, we have

$$m\frac{d^2 x_i'}{dt^2} = X_i.$$

As m is a scalar and $\frac{dx_i^2}{dt^2}$ is a vector, we deduce that X_i is also a vector.

3.2 Kinetic Energy and Potential Energy

If X_i be the force acting upon a particle whose position at any time t is x_i then the work done by the force is given by the line intergal $\int_C X_i dx_i$, the integral being taken along the path C of integration. Clearly the work as scalar. The scalar $\frac{1}{2}mv_i v_i$ is the kinetic energy of a particle whose mass is m and velocity, v_i.

3.3 Work Function and Potential Energy

The work done by a force as a particle moves from some given point a_i to another point x_i which depends in general the path followed from a_i to x_i and is not as such a point function. We shall however now consider a case where, the work function is necessary a scalar function.

Let now X_i be any field of force so that X_i is a vector point function. We say that this field is conservative if there exists a scalar point function U such that

$$\frac{\partial U}{\partial x_i} = U_i - X_i.$$

or in other words,

$$gradU = -X_i.$$

Thus, as a particle moves from a_i to another point x_i along any path C in the conservative field of force X_i the work done.

$$= \int_C X_i dx_i$$

$$= -\int_C U_i dx_i$$

$$= -\int_C \frac{\partial U}{\partial x_i} dx_i$$

$$-|U|_{a_i}^{x_i} = U(a_i) - U(x_i) \tag{3.2}$$

which is depends on the initial and final points and not on the path. We have

$$m\frac{d^2 x_i}{dt^2} = X_i,$$

$$m\frac{dx_i}{dt}\frac{d^2 x_i}{dt^2} = X_i\frac{dx_i}{dt}.$$

Integrating, we get

$$\int_{t_o}^{t} m\frac{dx_i}{dt}\frac{d^2x_i}{dt^2}dt = \int_{t_o}^{t} X_i\frac{dx_i}{dt}dt$$

$$\frac{1}{2}|m\frac{dx_i}{dt}\frac{dx_i}{dt}|t_o = \int_{t_o}^{t} X_idx_i. \qquad (3.3)$$

Writing $T = m\frac{dx_i}{dt}\frac{dx_i}{dt}$, we obtain from Eq.(3.2) and Eq.(3.3)

$$T(x_i) - T(a_i) = U(a_i) - U(x_i)$$

$$T(x_i) + U(x_i) = T(a_i) + U(a_i).$$

Thus, we see that for a conservative fields of force, the sum of the kinetic and potential energies remains constant.

3.4 Momentum and Angular Momentum

The momentum generally denoted by M_i of a system is the sum of the momenta of the particles of the system. Momentum of a system is also known as linear momentum. The angular momentum denoted by $H_i(O)$ or simply by H_i about any point O is the sum of the vector products of the position vectors relative to O and the particles of the sum of the moments of the momenta of the particle about O.

Thus, we have

$$M_i = \sum mv_i$$

$$H_i(O) = \sum \epsilon_{ikl}x_k mv_i,$$

where v_i is the velocity of the particle of mass m at x_i and the summation extends to all the particles of the system. The definition of moment as a skew-symmetric second-order tensor, in that, we would write

$$H_{ij}(O) = x_i v_j - x_j v_i.$$

The following discussion could also be carried out with $H_{ij}(O)$ instead of $H(O)$.

3.5 Moment of Inertia

We shall now introduce the concept of moment of inertia tensor of a rigid body relative to any point. It plays an important part in all discussions relative to the dynamics of rigid body. By the definition, of the moment of inertia of a material system about any line OL is

$$\sum mp^2$$

where p is the perpendicular distance of any particle of mass m of the system from the line OL, and the summation extends to all the particles of the system.

We take O as origin and any system of coordinate axes $OX_1X_2X_3$. Let l_i be the direction of cosines of any given line OL and let $P(x_i)$ be any particle of the system having mass m. We now need the length MP. Now the magnitude of the vector.

$$\epsilon_{ipq}l_pl_q$$

$$MP^2 = (\epsilon_{ipq}l_px_q)(\epsilon_{irs}l_rx_s)$$

$$= \epsilon_{ipq}\epsilon_{irs}l_pl_rx_qx_s$$

$$= l_r(x_qx_q\delta_{rs} - x_sx_r)l_s. \tag{3.4}$$

Thus, the moment of inertia of the system about the line OL

$$I = \sum ml_r(x_qx_q\delta_{rs} - x_sx_r)l_s$$

$$= l_r\sum(mx_qx_q\delta_{rs} - \sum mx_sx_r)l_s. \tag{3.5}$$

We write

$$l_{rs} = \sum m(x_qx_q\delta_{rs} - x_sx_r) \tag{3.6}$$

So that the moment of inertia about the given with direction cosines l_r is

$$l_rl_{rs}l_s.$$

It is shown that l_{rs} is a symmetric tensor of second order independent of the direction cosines of the lines OL and depending only on the configuration of the given material system relative to the given material system to the point O. It is known as inertia tensor. It may easily be seen that the matrix of the components of the tensor l_{rs} is

$$\begin{bmatrix} \sum m(x_2^2 + x_3^2) & -\sum mx_1x_2 & -\sum mx_1x_3 \\ -\sum mx_1x_2 & \sum m(x_3^2 + x_1^2) & -\sum mx_2x_3 \\ -\sum mx_3x_1 & -\sum mx_3x_2 & \sum m(x_1^2 + x_2^2) \end{bmatrix}.$$

In the case of a continuous system, sums have to be replaced by integrals. It may easily, seen that I_{11}, I_{22}, I_{33} are the moments of inertia about the axes OX_1, OX_2, OX_3 respectively. The product of inertia about the lines OX_1, OX_2 is $\sum mx_1x_2$ and the components I_{23}, I_{31}, I_{12} of the inertia tensor are the product of inertia with signs changed about the axes taken in pair.

3.6 Strain Tensor at Any Point

When some forces are applied to a body, the particle of the body undergoes relative displacement so that the body is deformed. We can say that the body has experienced a strain or that it is strained. We shall here be concerned about only small deformation. The rigid body displacement of translation and rotation do not produce any relative displacement of the particle so that such displacement do not constitute strain. We shall now proceed to the analysis strain produced in the neighbourhood of any given point of the body. Let the position of any particle of the body be determined by its coordinate x_i referred to any set of rectangular coordinate axes consider at any point $P(x_i)$ of the body and suppose that as a result of deformation it is displaced to another point $P'(x_i + s_i)$ so that, we have

$$\overrightarrow{PP'} = S_i.$$

clearly, S_i is a function of x_1, x_2, x_3 and so we may write

$$S_1 = f_1(x_1, x_2, x_3) \tag{3.7}$$

here we assume that the three-functions of f_1 possess continuous first-order partical derivatives.

We shall now compute the change in any vector

$$\overrightarrow{PQ} = h_i \tag{3.8}$$

which takes place as a result of the deformation of the body. Let Q be displaced to Q' so that the vector $\overrightarrow{PQ}$ becomes after displacement the vector $\overrightarrow{P'Q''}$. As the point Q is $(x_i + h_i)$, the displacement $\overrightarrow{QQ'}$ of the point Q is

$$f_1(x_1 + h_1, x_2 + h_2, x_3 + h_3)$$
$$= f_1(x_1, x_2, x_3) + h_1\frac{\partial f_1}{\partial x_1} + h_2\frac{\partial f_2}{\partial x_2} + h_3\frac{\partial f_3}{\partial x_3}$$
$$= f_1(x_1, x_2, x_3) + h_1\frac{\partial f_1}{\partial x_1}$$

where, we have neglected square and higher power of h_1, the point Q being in the neighbourhood of P point Q is

$$x_1 + h_1 + f_1(x_1, x_2, x_3) + h_1 \frac{\partial f_1}{\partial x_1} = x_1 + h_1 + h_1 \frac{\partial f_1}{\partial x_1}$$

Thus,

$$\overrightarrow{P'Q''} = (x_1 + h_1 + s_1 + h_j \frac{\partial f_i}{\partial x_j}) - (x_1 + s_1)$$

$$s_1 + h_j \frac{\partial f_i}{\partial x_j}$$

and

$$\overrightarrow{PQ} = h_i.$$

thus we see that a vector $\overrightarrow{PQ} = h_i$ is displaced to $\overrightarrow{P'Q''} = s_1 + h_j \frac{\partial f_i}{\partial x_j}$ so that the change in $\overrightarrow{PQ}$ is $h_j \frac{\partial f_i}{\partial x_j} = h_j a_{ij}$, say. By Quotient law, we see that a_{ij} is a tensor of order two. Its components being function of x_i, we now break up the deformation tendor a_{ij} as the sum of a symmetric and skew-symmetric tensor, we write

$$a_{ij} = \frac{1}{2}(a_{ij} + a_{ji}) + \frac{1}{2}(a_{ij} - a_{ji})$$

$$= e_{ij} + w_{ij}, \quad say$$

where, e_{ij} is symmetric and w_{ij} is skew symmetric h_i is displaced to $h_i + w_{ij}h_j$ as a result of rigid body motion of rotation about

P. The part of the displacement given by $e_{ij}h_i$ is called pure strain and the tensor e_{ij} is called the strain tensor at P. The pure strain gives the relative displacement of the particle as a result of the deformation.

3.7 Stress Tensor at any Point P

The force acting on a body are either external or internal. The force consists either of body force such as gravity that is acting on every particle. If F_1 denotes the body force vector per unit volume then the force acting on an element of volume ΔV is $F_1 \Delta V$. T_j denotes the surface force vector per unit area, then the force acting on an element of surface ΔS is $T_j \Delta S$. In order to discuss the internal force, we assume a small element of area ΔS inside the body and denotes the direction cosines of the normal to this element, which is approximately planer by n_j. We call one side of the element ΔS positive and other is negative. Then the action of the positive side on the negative side is the internal surface force $T_j \Delta S$, where T_j is the force per unit area on the element ΔS. It is called the stress tensor and in general. A function of the coordinate of the point, which determines the position of the element ΔS and of the direction cosines n_j of the normal to ΔS.

Consider a small rectangular parallelepiped with vertex at the point P, whose edges are parallel to the coordinate axes. We consider three stress vectors $T_{(1)i}$, $T_{(2)j}$, and $T_{(3)k}$ corresponding to the small elements of area through P, which are parallel to the coordinate planes. The stress vectors $T_{(i)j}$ will be called positive it act in the positive direction of the Y axes. If, however the external normal is co-directional with the negative Y axis. In other words, a stress which tends to stretch will be regarded as positive. we define nine quantities E_{ij} by the equation

$$E_{ij} = T_{(i)j} \tag{3.9}$$

We shall show that E_{ij} is a cartesian tensor called the stress Tensor. It follow the quotient law that E_{ij} is a cartesian tensor, we now cite several important case of stress.

3.7.1 Normal Stress

The vector T_{ij} is co-directional with $\pm n_j$. $E_{ij} = C\delta_{ij}$, where C is a constant hydrostatic pressure which, is an example of normal stress for which C is negative.

3.7.2 Simple Stress

Consider the stress tensor $E_{ij} = Cl_il_j$, where C is constant and l_i is a unit vector. Then the stress vector in the direction l_i is $T_j = Cl_j$. If C is negative, the stress is called a simple stress.

3.7.3 Shearing Stress

This is specified by the stress tensor $E_{ij} = C(l_im_j + l_jm_i)$, where C is a constant and l_i and m_i are unit vectors.

3.8 Generalised Hooke's Law

In the elementary theory of elasticity Hooke's law states that the tenson of a string is proportional to strain. The corresponding assumption in the general theory of elasticity is that the stress tensor is a linear homogeneous function of the strain tensor. That is

$$E_{ij} = \epsilon_{ijkl} e_{kl}. \tag{3.10}$$

It is follows from the quotient law that ϵ_{ijkl} is a cartesian tensor of the fourth order and it is called the elasticity tensor. Further, from the symmetry of E_{ij} and e_{kl}, we find that e_{ijkl} is symmetric not only with respect to the indices i and j but also with respect to k and l. A body is said to be homogenous. If the elastic properties of the body are independent of the point under consideration. We call a body isotropic if the elastic properties at a point are the same in all directions at that point. This means that the elasticity tensor, ϵ_{ijkl} itself under any rotation axes.

3.9 Isotropic Tensor

A cartesian tensor which transforms into itself under a rotation of axes is called isotropic tensor, we know two isotropic tensors namely δ_{ij} and ε_{ijk}. We shall now search for the most general isotropic tensor ϵ_{ijkl} of the fourth-order. Its transformation law becomes

$$\epsilon_{ijkl} = a_{ir} a_{js} a_{kl} a_{le} \epsilon_{rstv} \tag{3.11}$$

We can collect the given relation

$$\epsilon_{iiii} = \epsilon_{jjjj}$$

$$\epsilon_{iijj} = \epsilon_{iikk} = \epsilon_{lljj} = \epsilon_{iikk}.$$

$$\epsilon_{ijij} = \epsilon_{ikik} = \epsilon_{lllj} = \epsilon_{ikik}.$$

$$\epsilon_{ijji} = \epsilon_{ikki} = \epsilon_{ljjl} = \epsilon_{lkkl}.$$

Where i, j, k, l are unequal and the summation convention does not apply. All other components are zero. The most general solution of the equation is then

$$\epsilon_{ijkl} = \lambda\delta_{ij}\delta_{kl} + \mu\delta_{ik}\delta_{jl} + v\delta_{jl}\delta_{jk} + k\delta_{ij}\delta_{kl} \qquad (3.12)$$

where λ, μ, v and k are cartesian invariants and $\delta_{ijkl} = 1$. If all four indices are equal and otherwise zero. Finally we carry out a small rotation, which is represented by

$$a_{ik} = \delta_{ik} + s_{ik} \qquad (3.13)$$

where s_{ik} is skew-symmetric and of the first order in small quantities.

3.10 Exercises

(1) Show that the strain tensor satisfies the identical relation

$$\epsilon_{ijkl} + \epsilon_{klij} = \epsilon_{ikjk} + \epsilon_{jikl}$$

known as the equation of compatibility

(2) Show that the kinetic energy of a rigid body is

$$T = \frac{1}{2}Mu_iv_i + T_{ij}$$

where v_i is the velocity of the centre of a gravity. T_{ij} is the kinetic energy in regard to the motion of the body relative to the centre of gravity and M is the mass of the body.

(3) If the points of a rigid body have all two simultaneous velocity due to two angular velocity ω_r^1, ω_r^2. Show that the resultant velocity are given by an angular velocity vector

$$\omega_r = \omega_r^1 + \omega_r^2.$$

References

[1] Harold Jeeffreys (1931), Cartesian Tensor, PP(1-16), Combridge University Press (New York).

[2] David C. Kay, Theory and Problem of Tensor Calculus, PP(1-3), McGraw Hill, Washinton, D.C.

[3] Shanti Narayan (1961), Cartesian Tensor, PP(106-127), S. Chand, New Delhi.

[4] Barry Spain (1960), Tensor Calculus, PP(1-55), Dover Publication, Newyork.

[5] Zefer Ahson(2000). Tensor Analysis with Application, Anamaya Publisher, New Delhi.

4

Tensor in Analytic Solid Geometry

We have defined a vector analytically as a one suffix with prescribed transformation law, and we shall obtain a geometrical interpretation of the vector. This will bring our analytical treatment in line with the usual geometrical treatment. We shall also develop vector algebra in suffix notation corresponding to the usual algebra in term of one letter notation for vectors defined geometrically as directed line segments.

4.1 Vector as Directed Line Segments

Let a_i be the component of a vector relative to a system of axes $OX_1X_2X_3$. Take points A_1, A_2, A_3 in the three axes such that with some choice of scale,

$$OA_1 = a_1, OA_2 = a_2, OA_3 = a_3.$$

Let OA be the diagonal through O of the parallelopiped with OA_1, OA_2, OA_3 as coterminous edges. Then we say that the directed line segment select $\overrightarrow{OA}$ represents the vector a_i. Suppose that a_i and a_j are the components of a vector relative to two-system of axes $OX_1X_2X_3, O\bar{X}_1\bar{X}_2\bar{X}_3$ so that with the usual notation

$$\bar{a}_j = l_{ij}a_i. \tag{4.1}$$

Let $\overrightarrow{OA}$ be the directed line segment whose projection on the axes of the system $O\bar{X}_1\bar{X}_2\bar{X}_3$ is a_i. It will now be shown that the projection of $\overrightarrow{OA}$ on the axes of the system $O\bar{X}_1\bar{X}_2\bar{X}_3$ is $\bar{a}_j$.

Let l_i, $\bar{l}_j$ be the direction cosine of the line OA of the two systems of axes, so if r be the length of the line.

We have

$$a_i = rl_i, \tag{4.2}$$

and

$$\bar{l}_j = l_{ij}l_i,$$

or equivalently

$$l_i = l_{ij}\bar{l}_j,$$

now

$$\bar{a}_j = l_{ij}a_i$$

$$= l_{ij}rl_i$$

$$= rl_{ij}l_i$$

$$= rl_{ij}l_{ik}\bar{l}_k$$

$$= r\delta_{jk}\bar{l}_k$$

$$= r.\bar{l}_j.$$

Thus, the projection of $\overrightarrow{OA}$ on $O\bar{X}_j$ is $\bar{a}_j$.

4.2 Geometrical Interpretation of the Sum of two Vectors

Let the given vectors a_i, a_j be represented by two directed line segments $\overrightarrow{OA}, \overrightarrow{OB}$ and let OP be a diagonal of the parallelogram with OA, OB as a pair of adjacent sides. Then, we shall show that $\overrightarrow{OP}$ represents the vector $a_i + b_i$, this is equivalent to show that the projection of $\overrightarrow{OP}$ on any line is equal to the sum of the projections of $\overrightarrow{OA}$ and this is known to be true from elementary geometry.

4.3 Length and Angle between Two Vectors

The length of a vector is meant, to be the length of the directed line segment representing the vector. The angle between, the two vectors is meant the angle between the directed line segments representing the vectors. Let any vector a_i be represented by a directed line segment $\overrightarrow{OP}$. By the consideration of the rectangular parallelopiped with one diagonal OP and edges parallel to the coordinate axes, we may easily see that

$$OP = \sqrt{a_1^2 + a_2^2 + a_3^2} = \sqrt{a_1 a_1}.$$

Let θ be the angle between the directed line segments $\overrightarrow{OA}$ $\overrightarrow{OB}$ representing the vector a_1, b_1 from the ΔOAB, we have

$$\cos\theta = \frac{OA^2 + OB^2 - AB^2}{2OAOB}$$

$$= \frac{[a_1 a_1 + b_1 b_1 - (b_1 - a_1)(b_1 - a_1)]}{2\sqrt{a_1 a_1}\sqrt{b_1 b_1}}$$

4.4 Geometrical Interpretation of Scalar and Vector Products

If $\overrightarrow{OA}$, $\overrightarrow{OB}$ are two directed lie segments representing vectors a_i, a_j, then we have

$$a_i b = OA.OB.\cos\theta,$$

so that the scalar product of two vectors denotes the product of the length of the vectors and the cosine of the angle between the vectors. We now come to the case of the vector product

$$c_k = \epsilon_{ijk} a_i b_i,$$

of the vector a_i, b_j.

We first find the length of c_k, and we have

$$c_k c_k = (\epsilon_{ijk} a_i b_j)(\epsilon_{pqk} a_p b_p)$$

$$= \epsilon_{ijk}\epsilon_{pqk}a_i b_j a_p b_q$$

$$= (\delta_{ip}\delta_{jq} - \delta_{iq}\delta_{jp})a_i b_j a_p b_q$$

$$= (\delta_{ip}a_i b_p)(\delta_{jq}b_j b_q) - (\delta_{iq}a_i b_q)(\delta_{jp}b_j a_p)$$

$$= (a_p a_p)(b_q b_q) - (a_q a_q)(b_p b_p)$$

$$= OA^2 OB^2 - (OA.OB\cos\theta)^2$$

$$= (OA.OB\sin\theta)^2.$$

Thus, the length of c_k is $OA.OB\sin\theta$.

Again, we have

$$c_k a_k = \epsilon_{ijk}a_i b_j c_k = \begin{bmatrix} a_1 & a_2 & a_3 \\ a_2 & b_2 & b_3 \\ a_3 & c_3 & c_3 \end{bmatrix} = 0,$$

and similarly

$$c_k b_k = \epsilon_{ijk}a_i b_j b_k = 0.$$

Thus, the vector c_k is perpendicular to the both the vectors a_i, b_j.

Finally, we may see that the directed line segments $\overrightarrow{OA}$, $\overrightarrow{OB}$, $\overrightarrow{OC}$ representing the three vectors a_i, b_j, c_k from a right haded or left handed set according as the set of axes is right-handed or left handed. This follows from the fact that the determinant

$$\begin{bmatrix} a_1 & a_2 & a_3 \\ b_1 & b_2 & b_3 \\ a_2b_3 - a_3b_2 & a_3b_1 - a_1b_3 & a_1b_2 - a_2b_1 \end{bmatrix}$$

formed by the compoents of the three vectors in the order given is positive.

4.4.1 Scalar Triple Product

The scalar $\epsilon_{ijk}a_ib_jc_k$ is the scalar product of c_k with the vector product $\epsilon_{ijk}a_ib_j$ of a_i, b_j. The properties of the scalar triple product can all be easily shown. Consequences of the properties of the alternate tensor ϵ_{ijk}. In fact, they may not even the paid any special attention, if we work in suffix notation.

4.4.2 Vector Triple Products

The vector $\epsilon_{ijk}a_ib_j$ is the vector product of the vectors a_i, b_j and $\epsilon_{kpq}\epsilon_{ijk}a_ib_jc_p$ is the vector product of the vector of a_i, b_j with c_p. Here, i, j, k, p are dummy suffixes and q is the suffix. We now have

$$\epsilon_{kpq}\epsilon_{ijk}a_ib_jc_p = \epsilon_{pqk}\epsilon_{ijk}a_ib_jc_p$$

$$= (\delta_{pi}\delta_{qj} - \delta_{pj}\delta_{qi})a_ib_jc_p$$

$$= \delta_{pi}\delta_{qj}a_ib_jc_p - \delta_{pj}\delta_{qi}a_ib_jc_p$$

$$= (\delta_{pi}a_i)(\delta_{qj}b_j)c_p - (\delta_{pj}b_j)(\delta_{qi}a_i)c_p$$

$$= a_p b_q c_p - b_p a_p c_q$$

$$= a_p c_q b_q - b_p c_p a_q.$$

Here $a_p c_p$ *and* $b_p c_p$ are the scalar products of the vectors a_i, b_j and b_j, c_k respectively.

4.5 Tensor Formulation of Analytical Solid Geometry

We have seen that point and displacement are both cartesian tensors of order one. With the help of this fact, we shall briefly consider the tensorial formulation of linear analytic solid geometry.

4.5.1 Distance Between Two Points P(x_i) and Q(y_i)

We have

$$\overrightarrow{OP} = X_I, \quad \overrightarrow{OQ} = Y_I.$$

$$\overrightarrow{PQ} = \overrightarrow{OQ} - \overrightarrow{OP} = y_i - x_i,$$

$$PQ^2 = (y_i - x_i)(y_i - x_i).$$

4.5.2 Angle Between Two Lines with Direction Cosines

Let l_i, m_i are two points P, Q on the line through O parallel to the given lines and unit distance from O. Thus

$$\overrightarrow{OP} = l_I, \quad \overrightarrow{OQ} = m_i.$$

$$cos\theta = l_i m_i.$$

Where θ is the angle between two lines.

4.5.3 The Equation of Plane

Let l_i be the direction cosines of the normal to the given plane and p the length of the perpendicular from the origin to the plane. Take any point $P(x_i)$ on the plane. Let K be the foot of the perpendicular from the origin to the plane. The projection of OP on OK is OQ=p. Also

$$\overrightarrow{OP} = x_i$$

$$l_i x_i = p,$$

is the required equation. If the equation of a plane is

$$ax_i + d = 0.$$

the direction cosine of the normal to the plane are proportional to a_i and the length of the perpendicular from the origin to the plane is

$$\frac{|d|}{\sqrt{a_1 a_2}}.$$

4.5.4 Condition for Two Line Coplanar

Let two lines are $x_i = a_i + sl_i$, $\quad x_i = b_i + lm_i$. The line intersects, if and only if there exist scalar values s, t such that

$$a_i + sl_i = b_i + lm_i$$

$$a_i - b_i = tm_i - sl_i,$$

To eliminate s, t, we multiply with $\epsilon_{pq}l_im_k$ and also contract identifying i, p; j, q; k, r and obtain

$$\epsilon_{ijk}(a_i - b_i)l_jm_k = t\epsilon_{ijk}m_il_im_k - s\epsilon_{ijk}l_il_im_k.$$

Thus, the condition is

$$\epsilon_{ijk}(a_i - b_i)l_im_k = 0. \tag{4.3}$$

As x_i is any point on the second line, this condition can be written as

$$\epsilon_{ijk}(x_i - a_i)l_im_k = 0. \tag{4.4}$$

Thus, every point on the second line satisfied Eq.(4.4). Thus, Eq.(4.3) is the condition of coplanarity and assuming the condition to be satisfied, Eq.(4.4) is that of the plane through the two lines.

4.6 Exercises

(1) Show that rotations through 90° about the lines OX_1, OX_2 are given by

$$|a_{ij}| = \begin{pmatrix} 1 & 0 & 0 \\ 0 & 0 & -1 \\ 0 & 1 & 0 \end{pmatrix}, \qquad |b_{ij}| = \begin{pmatrix} 0 & 0 & 1 \\ 0 & 1 & 0 \\ -1 & 0 & 0 \end{pmatrix}$$

Show also that the resultants of these two rotations taken in the two orders are rotations through lines with direction cosines.

$$\frac{1}{\sqrt{3}}, \frac{1}{\sqrt{3}}, -\frac{1}{\sqrt{3}}; \frac{1}{\sqrt{3}}, \frac{1}{\sqrt{3}}, -\frac{1}{\sqrt{3}},$$

through angle 180°

(2) Verify that the tensor a_{ij}, which corresponds to a rotation through an angle θ about line l_i is orthogonal.

(3) Find the condition that the line $l_i x_i + p = 0$ may touch the quadric S.

References

[1] Harold Jeeffreys (1931), Cartesian Tensor, PP(16-23), Combridge University Press (New York).

[2] David C. Kay, Theory and Problem of Tensor Calculus, PP(1-3), McGraw Hill, Washinton, D.C.

[3] Shanti Narayan (1961), Cartesian Tensor, PP(54-80), S. Chand, New Delhi.

[4] Barry Spain (1960), Tensor Calculus, PP(17-23), Dover Publication, Newyork.

[5] Zefer Ahson(2000). Tensor Analysis with Application, Anamaya Publisher, New Delhi.

5

General Tensor

In this chapter, we will discuss the basic concepts of general tensor in three-dimensional Euclidean space. The notation of general tensor, related to arbitrary coordinate systems in curvilinear system of coordinates. In notation suffixes both as superscripts as well as subscripts and the coordinate of a point will be given by superscripts, according the coordinate of a point will be denoted by

$$x^1, x^2, x^3.$$

Here superscripts must on no account be confused. In this context, we shall also modified the summation convention and say that if in any symbol the same suffix appears both as a superscript as well as a subscript, then the symbol in question denotes a sum of three-numbers obtained by giving values 1,2,3 to the repeated suffix, for example

$$a_i^i = a_1^1 + a_2^2 + a_3^3,$$

$$a^i_{ji} = a^1_{j1} + a^2_{j2} + a^3_{j3}.$$

5.1 Curvilinear Coordinates

Consider arbitrary region R of space and three continuously differential functions

$$x^1 = f^1(y^1, y^2, y^3), x^2 = f^2(y^1, y^2, y^3), x^3 = f^3(y^1, y^2, y^3)$$

in general

$$x^i = f^i(y^1, y^2, y^3) \tag{5.1}$$

defined in R: y^1, y^2, y^3 is the rectangular Cartesian coordinates of any point p. Thus, to each point y^1, y^2, y^3 of R. The equation (5.1) associated a set of numbers x^1, x^2, x^3. We say that x^1, x^2, x^3 are the rectilinear coordinate of the point P.

5.2 Coordinate Transformation Equation

Consider now three equations

$$x^j = \phi^j(x^1, x^2, \bar{x^3}); j = 1, 2, 3. \tag{5.2}$$

where ϕ^j are three continuously differential functions. These equations associate to each point Q by x^1, x^2, x^3 another set of three numbers $\bar{x}^1, \bar{x}^2, \bar{x}^3$

Also we suppose that this association is one-one. No two different unbarred sets correspond to the same barred set. It is known from analysis that this fact will be guaranteed if the determinant

$$\left| \frac{\partial x^j}{\partial x^i} \right| \neq 0$$

$$\begin{vmatrix} \dfrac{\partial \bar{x}^1}{\partial x^1} & \dfrac{\partial \bar{x}^1}{\partial x^2} & \dfrac{\partial \bar{x}^1}{\partial x^3} \\[2ex] \dfrac{\partial \bar{x}^2}{\partial x^1} & \dfrac{\partial \bar{x}^2}{\partial x^1} & \dfrac{\partial \bar{x}^2}{\partial x^1} \\[2ex] \dfrac{\partial \bar{x}^3}{\partial x^1} & \dfrac{\partial \bar{x}^3}{\partial x^1} & \dfrac{\partial \bar{x}^3}{\partial x^1} \end{vmatrix} \neq 0$$

At any point P. In the following determinant is not zero. Thus, $\bar{x}^j$ can also be thought of as the curvilinear coordinate of the point P. The equation constitutes transformation equation between the two systems of coordinate x^i and $\bar{x}^j$.

5.3 Contravariant and Covariant Tensor

First, we shall consider two typical cases of transformation law, which will play an important role in the following discussion, consider two-systems of coordinate x^i and $\bar{x}^j$ and any point Q,

(1) we have

$$d\bar{x}^j = \frac{\partial \bar{x}^j}{\partial x^i} dx^i + \frac{\partial \bar{x}^j}{\partial x^2} dx^2 + \frac{\partial \bar{x}^j}{\partial x^3} dx^3$$

$$= \frac{\partial \bar{x}^j}{\partial x^i} dx^i. \tag{5.3}$$

Here, we have two sets dx^i and $d\bar{x}^j$ associated with the same point in the two system of coordinates x^i and $\bar{x}^j$ such that the equation (5.3) constitutes is the law of transformation from one set to the another. The coefficient of transformation depends only on the point Q and the transformation for the coordinate.

(2) Now any function ϕ of the point Q so that to each point P correspond a number ϕ can be thought of indifferently as a function of x^i as well as of $\bar{x}^j$ such that two sets x^i and $\bar{x}^j$ giving the coordinate of the same point correspond to the same value of ϕ, we have

$$\frac{\partial \phi}{\partial \bar{x}^j} = \frac{\partial \phi}{\partial x^1} \frac{\partial x^1}{\partial \bar{x}^j} + \frac{\partial \phi}{\partial x^2} \frac{\partial x^2}{\partial \bar{x}^j} + \frac{\partial \phi}{\partial x^3} \frac{\partial x^3}{\partial \bar{x}^j}$$

$$= \frac{\partial \phi}{\partial x^i} \frac{\partial x^i}{\partial \bar{x}^j}. \tag{5.4}$$

Here we have two sets $\frac{\partial \phi}{\partial x^i}$ and $\frac{\partial \phi}{\partial \bar{x}^j}$ associated with the same point in the two coordinates x^i and $\bar{x}^j$ such that the equation(5.49) constitute is the law of transformation from one set to the other, the coefficients of transformation depend only on the point and the equation of transformation of coordinates. In the two cases, it will be seen that the law of transformation are in general different in either case member of one set are expressed as linear combination of the member of other. The matrix of the coefficients of transformation is constant for different pairs of sets dx^i, $\dfrac{\partial \phi}{\partial x^i}$, $\dfrac{\partial \phi}{\partial x^j}$, provided only that we do not change the point Q.

5.4 Contravariant Vector or Contravariant Tensor of Order-One

Contravariant vector, if at each point Q, is represented by a set of three numbers such that if A^i, $\bar{A}_p$ of coordinate of two sets representing the entity relative to two systems of coordinates x^i, x^p, then

$$\bar{A}^p = \frac{\partial x^p}{\partial \bar{x}_i} A^i$$

Contravariant vector or Contravariant tensor of second order

$$A^{pq} = \frac{\partial \bar{x}^p}{\partial \bar{x}_i} \frac{\partial x^q}{\partial \bar{x}_j} A^{ij}$$

5.5 Covariant Vector or Covariant Tensor of Order-One

Covariant vector, if at each point Q it is represented by a set of three numbers such that if A^i, $\bar{A}_p$ of coordinate of two sets representing the entity to two system of coordinates x^i, x^p, then

$$\bar{A}_p = \frac{\partial x^i}{\partial \bar{x}^p} A_i.$$

It should be always noted that the upper suffix denotes contravariant and lower suffix denotes covariant.

Covariant vector or covariant tensor second order

$$A_{pq} = \frac{\partial \bar{x}^i}{\partial \bar{x}^p} \frac{\partial x^j}{\partial \bar{x}^q} A_{ij}.$$

5.6 Mixed Second-Order Tensor

A mixed second order tensor, if at each point Q, is represented by a set A_j^i, such that if A_j^i, $\bar{A}_q^p$ two sets representing the entity Relative to the two system of coordinate x^i, $\bar{x}^j$, then

$$\bar{A}_q^p = \frac{\partial \bar{x}^p}{\partial x^i} \frac{\partial x^i}{\partial \bar{x}^q} A_j^i.$$

The distinction between, contravariance and covariance does not exist in relation to the cartesian tensor.

5.7 General Tensor of Any Order

We can now easily define a tensor of any order with any number of contravariant and any number of covariant suffixes. For example, if we have an entity representable by sets A_{jk}^i, $\bar{A}_{qr}^p$ Relative to coordinate x^i, $\bar{x}^j$, such that

$$\bar{A}_{qr}^p = \frac{\partial \bar{x}^p}{\partial x^i} \frac{\partial x^i}{\partial \bar{x}^q} \frac{\partial x^k}{\partial \bar{x}^r} A_{jk}^i,$$

then we say that A_{jk}^i is a third-order tensor with one contravariant and two covariant suffixes.

5.8 Metric Tensor

Consider a rectangular cartesian system y^i and any curvilinear system x^i, consider two points P and Q near each other with coordinate in the two system.

$$y^i, y^i + dy^i$$

$$x^i, x^i + dx^i.$$

- Let ds denote the length PQ. We call ds the element of length or the line element, we have

$$(ds)^2 = dy^1 dy^1 + dy^2 dy^2 + dy^3 dy^3 = dy^i dy.^i \qquad (5.5)$$

Also

$$dy^i = \frac{\partial y^i}{\partial x^m} dx^m$$

$$dy^i dy^i = \frac{\partial y^i}{\partial x^m} dx^m \frac{\partial y^i}{\partial x^n} dx^n$$

$$= g_{mn} dx^m dx^n,$$

where

$$g_{mn} = \frac{\partial y^i}{\partial x^m} \frac{\partial y^i}{\partial x^n}.$$

Clearly,

$$g_{mn} = g_{nm}$$

Thus, we have

$$(ds)^2 = g_{mn}dx^m dx^n,$$

where $(ds)^2$ is invariant and dx^m is an arbitrary vector, it follows by applying quotient law twice that g_{mn} is a covariant second-order tensor. This is known as metric tensor for the space. It is easily seen that

$$g_{mn} = \begin{vmatrix} \dfrac{\partial y^i}{\partial x^i} & \dfrac{\partial y^2}{\partial x^i} & \dfrac{\partial y^3}{\partial x^i} & \dfrac{\partial y^i}{\partial x^i} & \dfrac{\partial y^i}{\partial x^2} & \dfrac{\partial y^i}{\partial x^3} \\ \dfrac{\partial y^i}{\partial x^2} & \dfrac{\partial y^2}{\partial x^2} & \dfrac{\partial y^3}{\partial x^2} & \dfrac{\partial y^2}{\partial x^i} & \dfrac{\partial y^2}{\partial x^2} & \dfrac{\partial y^2}{\partial x^3} \\ \dfrac{\partial y^i}{\partial x^3} & \dfrac{\partial y^2}{\partial x^3} & \dfrac{\partial y^3}{\partial x^3} & \dfrac{\partial y^3}{\partial x^i} & \dfrac{\partial y^3}{\partial x^2} & \dfrac{\partial 3}{\partial x^3} \end{vmatrix}$$

$$= \left| \frac{\partial y^i}{\partial x^i} \right|^2 \neq 0. \tag{5.6}$$

It is important to remember that the tensor g_{mn} is essentially a characteristic of the Euclidean space which will have different components for different coordinate system to which the space may be referred.

5.9 Associate Contravariant Metric Tensor

We denote by g^{mn} the cofactor of g_{mn} in the matrix $[g_{mn}]$, divided by the determinant of the matrix. We have

$$g^{mn}g_{np} = \delta^m_p.$$

As g_{np} is not an arbitrary covariant tensor, we cannot apply Quotient law to show that g^{mn} is a contravariant second-order tensor. Let u^p be any arbitary contravariant vector, then

$$g_{np}u^p = v_n,$$

is also an arbitrary covariant vector. We have

$$g^{mn}v_n = g^{mn}g_{np}u^p = \delta_p^m u^p = u^m.$$

Here v_n is any arbitary convariant vector and u^m is also a vector, hence by quotiant law g^{mn} is a contravariant second-order tensor. The tensor g^{mn} may be called contravariant metric tensor.

5.10 Associate Metric Tensor

Raising and lowering of suffixes, we shall discuss the important process of raising and lowering the suffixes of a tensor in order to obtain. New tensor associated with those given. Let now u^r be any contravariant vector, then $g_{mr}u^r$ is a covariant vector. We say that this vector has been obtained on lowering the index and denote it again by the symbol u_m. Thus, we write

$$u_m = g_{mr}u_r.$$

Similarly, the contravariant vector

$$v^m = g_{mr}u_r$$

is said to be obtained by raising the index. The process of lowering and raising of suffixes can be applied in a tensor of every type. For example, consider the tensor

$$u^{ijk}{}_{pq}$$

and suppose that we wish to lower j. This is given on inner multiplication with g_{mj}. Thus, the requried tensor is

$$g_{mj}u^{ijk}{}_{pq}.$$

Two tensors are said to be Associated. If either is obtained from the other by any combination of the process of raising and lowering the suffixes. It is important to notice that the sets of components of two associated vector Relative to any cartesian rectangular system of coordinates are the same. This if u^n be the components of a contravariant vectors relative to a cartesian rectangular system, the components of the associated covariant vector relatively to the same system are

$$\delta_{mn}u^n = u_m$$

Thus, the sets of components are the same.

5.11 Christoffel Symbols of the First and Second -Kind

Consider any Curvilinear system of coordinate and any curve so that the coordinate x^i of any point on the curve is given as a function of t.

If u^i and u^j be the components of a vector at any point of the curve Relative to the two systems of axes, we have

$$u^i = \frac{\partial x^i}{\partial x^m} u^m. \tag{5.7}$$

Here, u^i, u^m, x^i, x^m are all functions of t. Suppose that the set of vectors associated with different points of the curve are equal and parallel so that u^i are the constant.

Differentiating Eq.(5.7) with respect to t, we get

$$0 = \frac{du}{dt} = \frac{\partial^2 x^i}{\partial x^m \partial x^n} \frac{dx^n}{dt} u^m + \frac{\partial x^i}{\partial x^m} \frac{du^m}{dt}.$$

On inner multiplication with $g^{rp} \frac{\partial x^i}{\partial x^p}$, we get

$$\frac{du^r}{dt} + g^{rp} \frac{\partial^2 x^i}{\partial x^m \partial x^n} \frac{\partial x^i}{\partial x^p} \frac{dx^n}{dt} u^m = 0. \tag{5.8}$$

Here, the summation for i=1,2,3 is implied even through the repeated suffix of i at the same level. We have

$$g^{rp} \frac{\partial x^i}{\partial x^p} \frac{\partial x^i}{\partial x^m} \frac{du^m}{dt} = g^r g_{mp} \frac{du^m}{dt} = \delta^r_m \frac{du^r}{dt}.$$

Eq.(5.8) gives the condition to be satisfied by the set of equal and parallel vectors u^j along a given curve. We saw now in Eq.(5.1) to an important information by introducing what are known as christoffel symbols of the first and second kind. In this connection, we consider the following expression in Eq.(5.1)

$$\frac{\partial^2 x^i}{\partial x^m \partial x^n} \frac{\partial x^i}{\partial x^p}.$$

Now, we have

$$g_{mn} = \frac{\partial x^i}{\partial x^m} \frac{\partial x^i}{\partial x^n}$$

$$\frac{\partial g_{mn}}{\partial x^p} = \frac{\partial^2 x^i}{\partial x^m \partial x^p} \frac{\partial x^i}{\partial x^n} + \frac{\partial^2 x^i}{\partial x^n \partial x^p} \frac{\partial x^i}{\partial x^m}. \tag{5.9}$$

Interchanging m, n, p cyclically in Eq.(5.9), we get

$$\frac{\partial g}{\partial x^m} = \frac{\partial^2 x^i}{\partial x^n \partial x^m} \frac{\partial x^i}{\partial x^p} + \frac{\partial^2 x^i}{\partial x^p \partial x^m} \frac{\partial x^i}{\partial x^n} \tag{5.10}$$

$$\frac{\partial g_{pm}}{\partial x^n} = \frac{\partial^2 x^i}{\partial x^p \partial x^n} \frac{\partial x^i}{\partial x^m} + \frac{\partial^2 x^i}{\partial x^m \partial x^n} \frac{\partial x^i}{\partial x^p}. \tag{5.11}$$

Equations (5.9),(5.10) and (5.11), we get

$$\frac{\partial g_{np}}{\partial x^m} + \frac{\partial g_{pm}}{\partial x^n} - \frac{\partial g_{mn}}{\partial x} = 2\frac{\partial^2 x^i}{\partial x^m \partial x^n} \frac{\partial x^i}{\partial x^p}. \tag{5.12}$$

and we write

$$[mn, p] = \frac{\partial g_{np}}{\partial x^m} + \frac{\partial g_{pm}}{\partial x^n} - \frac{\partial g_{mn}}{\partial x}. \tag{5.13}$$

Thus, with the help of Eq.(5.12) and Eq.(5.13), we re-write Eq.(5.1) as

$$\frac{du^r}{dt} + g^{rp}[mn, p] u^m \frac{dx^n}{dt} = 0.$$

Again writing

$$g^{rp}\,[mn, p] = \begin{bmatrix} r \\ mn \end{bmatrix}$$

we obtained

$$\frac{du^r}{dt} + \begin{bmatrix} r \\ mn \end{bmatrix} u^m \frac{dx^n}{dt} = 0 \qquad (5.14)$$

as the required form of condition, if u^r is a set of equal and parallel vectors along the curve. The symbols $[mn, p]$ and $\begin{bmatrix} r \\ mn \end{bmatrix}$ are known as christoffel symbols of the first and second kind respectively. Both symbols are easily seen to be symmetrical in m, n.

5.12 Covariant Derivative of a Covariant Vector

Let X be any covariant vector field, we take a parallel vector field u^r so that u^r satisfied the equation

$$\frac{du^r}{dx^m} + \begin{bmatrix} r \\ mn \end{bmatrix} u^m = 0. \qquad (5.15)$$

Then $x_r u^r$ is a scalar, $\dfrac{\partial(x_r u^r)}{\partial x^i}$ is a covariant vector, now

$$\frac{\partial(x_r u^r)}{\partial x^i} = x_r \frac{\partial u^r}{\partial x^i} + u^r \frac{\partial x_r}{\partial x^i}. \qquad (5.16)$$

From Eq.(5.15)

$$\frac{\partial u^r}{\partial x^i} = - \begin{bmatrix} r \\ mn \end{bmatrix} u^m. \tag{5.17}$$

From Eq. (5.16) and Eq.(5.17), we have

$$\frac{\partial (x_r u^r)}{\partial x^i} = x_r \begin{bmatrix} r \\ mn \end{bmatrix} u^m + u^r \frac{\partial x_r}{\partial x^i}$$

$$= \frac{\partial x^m}{\partial x^i} - x_r \begin{bmatrix} r \\ ms \end{bmatrix} u^m.$$

Here, u^m is an arbitrary contravariant vector and $\dfrac{\partial (x_r u^r)}{\partial x^i}$ is a covariant vector. Thus

$$\frac{\partial x_m}{\partial x_i} - \begin{bmatrix} r \\ ms \end{bmatrix} x_r$$

is a covariant second-order tensor called the covariant derivative of x_r, we write

$$x_{mrs} = \frac{\partial x_m}{\partial x_i} - \begin{bmatrix} r \\ ms \end{bmatrix} x_r.$$

5.13 Covariant Derivative of a Contravariant Vector

If x^r any contravariant vector, then it can be shown that

$$\frac{\partial x^r}{\partial x^i} + \begin{bmatrix} r \\ ms \end{bmatrix} x^m$$

is a mixed second-order tensor. It is called the covariant derivative of x^r and we write

$$x^r_{ms} = \frac{\partial x^r}{\partial x^s} + \begin{bmatrix} r \\ ms \end{bmatrix} x^m.$$

This can be proved by taking a contravariant vector field x^r along a curve C and parallel field of covariant vector u_r along C and considering the invariance of $x^r u_r$.

Thus, we have the two following important results:

$$X_{mrs} = \frac{\partial x_m}{\partial x^s} - \begin{bmatrix} r \\ ms \end{bmatrix} X_{rs} \tag{5.18}$$

$$X^r_{ms} = \frac{\partial x_r}{\partial x^s} + \begin{bmatrix} r \\ ms \end{bmatrix} X^m. \tag{5.19}$$

5.14 Exercises

(1) Find the components of the Christoffell symbols for the metric

$$ds^2 = r^2 d\theta^2 + r^2 \sin^2\theta d\phi^2$$

(2) Compute all the Christoffell symbols for a space whose line element is given by

$$ds^2 = -e^\lambda dr^2 - r^2 d\theta^2 - r^2 \sin^2\theta d\phi^2 + e^\mu dt^2$$

(3) The length ds of a line element in a two-dimensional surface θ, ϕ is given by

$$ds^2 = R^2 d\theta^2 + R^2 sin^2\theta d\phi^2$$

where R is constant. Find all the components of the metric tensor $g_{\mu\nu}$ and the christoffel symbols of first kind for this surface.

(4) Prove that

$$g_{ij,k} = [ik, l] + [jk, i].$$

(5) Compute the christoffel symbols of the first and the second kind in a plane in terms of polar coordinate r, θ.

(6) If A^{ijk} is a skew-symmetric tensor, show that

$$A^{ijk} \left\{ \begin{array}{c} i \\ ij \end{array} \right\} = A^{ijk} \left\{ \begin{array}{c} l \\ jk \end{array} \right\} = A^{ijk} \left\{ \begin{array}{c} l \\ ik \end{array} \right\}.$$

References

[1] Harold Jeeffreys (1931), Cartesian Tensor, PP(43-57), Combridge University Press (New York).

[2] David C. Kay, Theory and Problem of Tensor Calculus, PP(1-3), McGraw Hill, Washinton, D.C.

[3] Shanti Narayan (1961), Cartesian Tensor, PP(129-154), S. Chand, New Delhi.

[4] Barry Spain (1960), Tensor Calculus, PP(27-37), Dover Publication, Newyork.

[5] Zefer Ahson(2000). Tensor Analysis with Application, Anamaya Publisher, New Delhi.

[6] U.C. De (2010) Tensor Calculus. PP(62-135), Narosa Publishing House, New Delhi.

6

Tensor in Relativity

6.1 Special Theory of Relativity

In classical mechanics, the position of a point in a space at which any point P can be determine by its three space coordinate x^1, x^2, x^3 referred to some rectangular cartesian system. Also an observer can measured the time t at which the event takes place by mean of clock, space and time cordinate comprised of the four numbers x^1, x^2, x^3 and t.

Einstein examined the concept of simultaneity and comes to the conclusion that simultaneous events at different points has no meaning without qualification. Einstein arrived at the special theory of relativity which he based on the two principles.

(1) It is impossible to detect the unaccelerated translating motion of a system through space.

(2) The velocity c of a ray of light is a constant, which does not depend on the relative velocity of its source and the observer. Let us assume two systems S ans S, which coincide at the

time t=0, such that S, moves with constant velocity v along x axis of the system. Then the Lorentz transformation, which can be deduced from the two principles of special relativity, connects the space coordinates and the time of both system by the equation

$$x^{1,} = \beta(x^1 - vt), x^{,2} = x^2, x^{,3} = x^3, t^. = \left(1 - \frac{vx'}{c^2}\right)$$

where $\beta = (1 - \frac{v^2}{c^2})^{\frac{1}{2}}$.

We can verify that

$$- (dx^{.1})^2 - (dx^{.2})^2 - (dx^{.3})^2 + c^2(dt^.)^2$$

$$= (dx^1)^2 - (dx^2)^2 - (dx^3)^2 + c^2(dt^.)^2. \tag{6.1}$$

The invariance of this equation with respect to lorentz transformation suggests that the Minkowski space defined by the metric

$$d\sigma^2 = -(dx^1)^2 - (dx^2)^2 - (dx^3)^2 + c^2(dx^4)^2 \tag{6.2}$$

where we have written $x^4 = t$ is approximate for the geometrical discussion of special relativity. We denote the line element of this four dimensional space by $d\sigma$ is the physical distance between two near points. The velocity u of a particle, which is at the point x^i has the components $u^i = \frac{dx^i}{dt}$ referred to the system S. It follows that

$$\frac{dx^i}{dt} = \frac{1}{c}(1 - \frac{v^2}{c^2})^{-\frac{1}{2}}. \tag{6.3}$$

The four-dimensional Minkowski momentum vector is defined by $m_o c \frac{dx^\alpha}{d\sigma}$, where m_o is a constant. The special theory identifies the fourth components $m_o c \frac{dx^4}{d\sigma}$ with the mass m of the moving particle. We have

$$m = m_0(1 - \frac{v^2}{c^2})^{-\frac{1}{2}} \tag{6.4}$$

the constant m_o is the mass when u=0 and it is called the rest mass of the particle. The mass m, which clearly increase the velocity is called the relativistic mass of the particle. The components

$$m_o c \frac{dx^i}{d\sigma} = m_o c \frac{dx^i}{dt} \frac{dt}{d\sigma} = m \frac{dx^i}{dt}$$

are the generalization of the newtonion momentum vector. We defined the fourth dimensional Minikowaski momentum vector F^α by

$$F^\alpha = m_o c^2 \frac{d^2 x^\alpha}{d\sigma^2}$$

$$= c^2 \frac{d}{d\sigma}(m_o \frac{dx^\alpha}{d\sigma})$$

$$= (1 - \frac{v^2}{c^2})^{-\frac{1}{2}} \frac{d}{dt}(m_o \frac{dx^\alpha}{dt}). \tag{6.5}$$

The Newtonian force vector is

$$x^i = \frac{d}{dt}\left(m_o \frac{dx^i}{dt}\right), F^i = \left(1 - \frac{v^2}{c^2}\right)^{-\frac{1}{2}} x^i.$$

The motion of a particle which moves under the action of some force system can be represented in Minkowski space by a curve called the world line of the particle. If no force acts on the particle, we have $\frac{d^2 x^\alpha}{d\sigma^2} = 0$, thus the world line of a free particle is a geodesic of the minkowski space.

The velocity of a light ray is the constant c. In order to discuss the mechanics of a continuous medium, we introduce the symmetrical four dimensional energy momentum Tensor $T^{\alpha\beta}$ defined by

$$T^{ij} = T^{ji} = \rho u^i u^j - E^{ij}; T^{i4} = T^{4i} = \rho u^i; T^{44} = \rho$$

where ρ is the density and E^{ij} is the cartesian stress tensor. If we change to spherical polar coordinate r, θ and ϕ the metric of minkowski space becomes

$$d\sigma^2 = -dr^2 - r^2 d\theta^2 - r^2 \sin^2\theta d\phi^2 + c^2 dt^2 \qquad (6.6)$$

6.2 Four-Vectors in Relativity

Four-vectors leading to two different types contravariant and covariant four vector. Let us consider a four-vector A. It has four component, one time A_l and three components A_x, A_y, A_z. In the four-dimensional real Minikowski space, because of the introduction of the metric tensor $g_{\mu v}$, complications arise. There is two types of four-vectors, a contravariant four vector denoted by a suprescript

$$A^\mu; A^v, A^1, A^2, A^3 = A_t, A_x, A_y, A_z.$$

and a covariant four vector denoted by a subscript

$$A^\mu; A_o, A_1, A_2, A_3 = A_t, -A_x, -A_y, -A_z.$$

The metric tensor is a 4×4 diagonal matrix

$$g_{\mu v} = \begin{bmatrix} 1 & 0 & 0 & 0 \\ 0 & 1 & 0 & 0 \\ 0 & 0 & -1 & 0 \\ 0 & 0 & 0 & -1 \end{bmatrix}. \tag{6.7}$$

It can be written in conscise form as

$$g_{\mu v} = g^{\mu v} = \begin{cases} 0 & \mu \neq v \\ 1 & \mu = v = 0 \\ -1 & \mu = v = 1, 2, 3 \end{cases}. \tag{6.8}$$

Using the metric tensor, one can convert a contravariant four vector into a covariant four vector and vice versa.

$$A_\mu = \sum_v g_{\mu v} A^v, \qquad \mu = 0, 1, 2, 3. \qquad (6.9)$$

$$A^\mu = \sum_v g^{\mu v} A_v, \qquad \mu = 0, 1, 2, 3. \qquad (6.10)$$

If we use Einstein's convention that repeated index involves the summation symbol that we used above. The scalar product of any two four-vector A and B can be written as

$$A.B = A^\mu B_\mu = A^0 B_0 + A^1 B_1 + A^2 B_2 + A^3 B_3$$

$$= A_t B_t - A_x B_x - A_y B_y - A_z B_z$$

$$A_t B_t - A.B. \qquad (6.11)$$

The differential operators in the four-dimensional space time x^μ transforms as the components of a covariant four-vector.

$$\frac{\partial}{\partial x^\mu} = \frac{\partial x^\mu \partial}{\partial x'^\mu \partial x^\mu} \qquad (6.12)$$

$$\partial_\mu = \frac{\partial}{\partial x^\mu} = \left(\frac{\partial}{\partial x^\mu} \nabla \right). \qquad (6.13)$$

The corresponding contravariant four vector is

$$\partial^\mu = g^{\mu v} \partial_v = \left(\frac{\partial}{\partial x_0} \nabla \right). \qquad (6.14)$$

It is clear that $g_{\mu v}$ contains all the information about the geometry of the space. In the case of Minkowski's space time. If we confine the special theory of relativity the metric tensor $g_{\mu\nu}$ plays a passive role but it play an active role in general relativity since the space time geometry is not fixed in advance and can be curved depending on the distribution of matter. So in special relativity one can avoid the distinction between the contravariant and covariant vectrors and define the scalar product of four vectors by

$$A.B = A_o B_0 - A.B = A_o B_o - A_x B_x - A_y B_y - A_z B_z \quad (6.15)$$

6.3 Maxwell's Equations

The classical theory of electrodynamics according to Lorentz is specified by the electric potential ϕ, which is scalar and the magnetic potential A_i, which is vector. Electric field strength vector E_i and the magnetic vector H_i are derived from there-potential by the equation

$$E_i = -grad\phi - \frac{1\delta A}{c\delta t} \quad (6.16)$$

$$H_i = curl A_i. \quad (6.17)$$

using electrostatic units, maxwell's equations are

$$div E_i = 4\pi\rho \quad (6.18)$$

$$div H_i = 0 \quad (6.19)$$

$$curl E_i + \frac{1\delta H_i}{c\delta t} = 0 \quad (6.20)$$

$$curl H_i - \frac{1\delta E_i}{c\delta t} = \frac{4\pi J_i}{c} \tag{6.21}$$

where J_i is the current density vector and ρ is the charge density. In Minkowski space, let us form the four-dimension potential vector ϕ_α and the four-dimension current density vector J^α defined respectively by

$$\phi_\alpha = (-A_1, -A_2, -A_3, c\phi)$$

$$J_\alpha = (-J_1, -J_2, -J_3, \rho)$$

with respect to a particular coordinate system. Next we introduce the skew symmetric tensor

$$\eta_{\alpha\beta} = \phi_{\alpha\beta} - \phi_{\beta\alpha}$$

$$= \frac{\delta\phi_\alpha}{\delta x_\beta} - \frac{\delta\phi_\beta}{\delta x_\alpha}$$

and we immediately calculate that its non-vanishing components in the given coordinate system are

$$\eta_{23} = -\eta_{32} = H_1; \eta_{31} = -\eta_{13} = H_2; \eta_{12} = -\eta_{21} = H_3;$$

$$\eta_{14} = -\eta_{41} = cE_1; \eta_{24} = -\eta_{42} = cE_2; \eta_{34} = -\eta_{43} = cE_3;$$

The non-vanishing contravariant components $\eta^{\alpha\beta}$ may now be obtained and are

$$\eta^{23} = -\eta^{32} = H_1; \eta^{31} = -\eta^{13} = H_2; \eta^{12} = -\eta^{21} = H_3;$$

$$\eta^{14} = -\eta^{41} = \frac{-E_1}{c}; \eta^{24} = -\eta^{42} = \frac{E_2}{c}; \eta^{34} = -\eta^{43} = \frac{E_3}{c};$$

We now write Maxwell's equation in terms of η and J and the results are readily verified to be respectively

$$\frac{\partial \eta^{41}}{\partial x^1} + \frac{\partial \eta^{42}}{\partial x^2} + \frac{\partial \eta^{43}}{\partial x^3} = \frac{4\pi}{c} J^4$$

$$\frac{\partial \eta_{23}}{\partial x^1} + \frac{\partial \eta_{31}}{\partial x^2} + \frac{\partial \eta_{12}}{\partial x^3} = 0$$

$$\frac{\partial \eta_{1j}}{\partial x^4} + \frac{\partial \eta_{j4}}{\partial x^i} + \frac{\partial \eta_{4j}}{\partial x^j} = 0$$

$$\frac{\partial \eta^{i1}}{\partial x^1} + \frac{\partial \eta^{i2}}{\partial x^2} + \frac{\partial \eta^{i3}}{\partial x^3} + \frac{\partial \eta^{i4}}{\partial x^4} = \frac{4\pi}{c} J^i.$$

The first and last of these equations combine together into the form

$$\eta^{\alpha\beta}_{\beta} = \frac{4\pi}{c} J^\alpha \tag{6.22}$$

$$\eta_{\zeta\beta,\gamma} + \eta_{\beta\gamma,\alpha} + \eta_{\gamma\alpha,\beta} = 0. \tag{6.23}$$

we have accordingly written, maxwell's equation in tensor form in minkowski space. Thus, the are invariant under the Lorentz group of transformations.

6.4 General Theory of Relativity

General theory of relativity, which was developed by Einstein in order to discuss gravitation, he postulated the principle of covariance, which shows that the law of physics must be independent of the space time coordinates. This swept away the privileged role of the Lorentz transformation. As a result, Minkowski space was replaced by the Riemannian V4 with the general metric

$$d\sigma^2 = g_{\alpha\beta}dx^\alpha dx^\beta. \tag{6.24}$$

Einstein also introduced the principle of equivalence, which in essence states that the fundamental tensor $g_{\alpha\beta}$ can be chosen to account for the presence of a gravitational field. That is $g_{\alpha\beta}$ depends on the distribution of matter and energy in physical space. Matter and energy can be specified by the energy momentum tensor $T^{\alpha\beta}$, which in the special theory satisfied the equation $T_\alpha^{\alpha\beta} = F^\beta$. Einstein's tensor defined by

$$G_\beta^\alpha = g^{\alpha\gamma}R_{\beta\gamma} - \frac{1}{2}R\delta_\beta^\alpha \tag{6.25}$$

Satisfies the equation $G_{\beta\alpha}^\alpha = 0$. This equation of motion requires $T_{\beta\alpha}^\alpha = 0$, but very remarkably $G_{\beta\alpha}^\alpha = 0$. is an identity in Riemannian geometry. This led Einstein to propose the relation

$$kT_\beta^\alpha + G_\beta^\alpha = 0. \tag{6.26}$$

In effect, these equations form the link between the physical energy momentum tensor T^α_β and the geometerical tensor G^α_β of the V4 of general relativity. In the special theory , the world line of free particles and of light rays are respectively the geodesics and the null geodesics of the Minkowski space. The principle of equivalance demands that all particles be regarded as free particle, when gravitation is the only force under consideration. Then it follows from the principle of covariance that the world line of a particle under the action of gravitational forces is geodesic of the V4 with the metric.

6.5 Spherically Symmetrical Metric

General relativity discusses several important problems in which the coordinate systems r, θ, ϕ and t are such that the metric takes the form

$$d\sigma^2 = -e^\lambda dr^2 - r^2 d\theta^2 - r^2 sin^2\theta d^2\phi + c^2 e^v d^2 t \qquad (6.27)$$

where λ and v are functions of r. A metric of this type is said to be spherically symmetrical. It is a generalization of the special relativity metric is expressed in spherical polars. The coefficients of dr^2 and dt^2 have been selected as exponential in order to ensure that the signature of $d\sigma^2$ is -2. Let us write $x^1 = r$, $x^2 = \theta$, $x^3 = \phi$ and $x^4 = ct$. The non zero components of the fundamental tensor are

$$g_{11} = -e^\lambda, \quad g_{22} = -r^2, \quad g_{33} = -r^2 sin^2\theta$$

and $g_{44} = e^v$.

The determinant g becomes

$$g = -r^4 sin^2\theta e^{\lambda tv}$$

and hence the non zero components of the conjugate symmetric fundamental tensor are

$$g^{11} = -e^{-\lambda}, \; g^{22} = -\frac{1}{r^2}, \; g^{33} = -\frac{1}{r^2 sin^2\theta}$$

and $g^{44} = e^{-v}$.

We obtained the components of the Einstein tensor for the spherically metric

$$G_1^1 = -\frac{1}{r^2} + e^{-\lambda}\left\{\frac{1}{r^2} + \frac{1}{r}v'\right\}, \tag{6.28}$$

$$G_2^2 = G_3^3 = e^{-\lambda}\left\{-\frac{1}{2r}\lambda - \frac{1}{4}\lambda'v' + \frac{1}{2}v'' + \frac{1}{2r}v' + \frac{1}{4}v'^2\right\}, \tag{6.29}$$

$$G_4^4 = -\frac{1}{r^2} + e^{-\lambda}\left\{\frac{1}{r^2} - \frac{1}{r}\lambda'\right\}, \tag{6.30}$$

$$G_\beta^\alpha = 0 \quad for \; \alpha \neq \beta \tag{6.31}$$

6.6 Planetary Motion

Let us investigate the motion of a planet in the gravitational field of the sun. The sun will be selected as a gravitational particle and the planet as a free particle whose mass is so small that it does not affect the metric differential equation. for the planet of our solar system is

$$\frac{d^2u}{d\phi^2} + u = \frac{m}{c^2h^2} + \frac{3mu^2}{c^2}.$$

(6.32)

The term $\frac{m}{c^2h^2}$ is much larger than $\frac{3mu^2}{c^2}$, but when, we neglected this latter term, we obtained Newton's equation for the motion of a planet. Thus the first approximation to the solution of Eq.(6.32) is the Newtonian solution is

$$u = \frac{m}{c^2h^2}(1 + e\cos(\phi - \epsilon),$$

where e is the eccentricity of the elliptic orbit and ϵ is the longitude of perihelion. A second approximation to the solution can then be obtained in the form

$$u = \frac{m}{c^2h^2}(1 + e\cos(\phi - \epsilon - \Delta\epsilon),$$

where

$$\Delta\epsilon = \frac{3m^2\phi}{c^4h^2}$$

This means that the major axes of the elliptic orbit is slowly rotating about its focus (sun).

6.7 Exercises

(1) Show that (a) E.B and (b) $E^2 - B^2$ are invariant under Lorentz transformation.

(2) Prove that the curvature invariant of Einstein's universe is $R = 6/R''^2$.

References

[1] J.A. Schouten, Der Ricci-Kalkul, (1954).

[2] Barry Spain, Tensor Calculus, (1952).

[3] H. Weyl, Space Time Matter, (1922).

[4] I.S. Sakolinkoff, Mathmatical Theory of Elasticity, (1946).

[5] A. Einstein, Theory of Relativity, (1958).

[6] W. Pauli, Theory of Relativity, (1958).

[7] I.S. Sakolinkof, Tensor Analysis, (1967).

[8] A.W. Joshi, Matrices and Tensor in Physics, (1975).

7

Geodesics and Its Coordinate

7.1 Families of Curves

Family of curves given by $y = f(x,t)$. For every t, we obtain a particular curve, if t is fixed and x varies using the symbol d to represent the differential. If we keep x is fixed and that u varies. Then the symbol δ respresent the differentials and this differential will be a variation. The commutative law of partial derivatives

$$\phi_{xt} = \phi_{tx}, \tag{7.1}$$

now

$$d\delta\phi = d(\phi_t\delta t) = (d\delta t)\delta t = (\phi_{tx}dx)\delta t = \phi_{tx}dx\delta t$$

$$\delta d\phi = \delta(\phi_x dx) = (\delta\phi_x)dx = (\phi_{xt}\delta t)dx = \phi_{xt}\delta t dx.$$

Hence,

$$d\delta\phi = \delta d\phi.$$

7.2 Euler's Form

Let $A(x_1, y_1)$ and $B(x_2, y_2)$ are two points. This is a particular curve for a family, if the curve

$$y = y(x, t) \quad for \quad t = 0.$$

If all the curves have the same extremities A and B, then we have $\delta y = 0$. If δt is the small values of neighbouring curves, then by Leibnitz's rule

$$\int_{x_1}^{x_2} (\delta\phi) dx = 0. \tag{7.2}$$

now

$$\delta\phi = \phi_x \delta x + \phi_y \delta y + \phi_{y'} \delta y' = \phi_y \delta y + \phi_{y'} \delta y'.$$

since $\delta x = 0$ and $\delta y = 0$ at A and B

$$\int_{x_1}^{x_2} (\phi_y \delta y + \phi_{y'} \delta y') \delta x = 0.$$

Euler's condition

$$= \phi_y - \frac{d}{dx}\phi_{y'}.$$

7.3 Geodesics

n-dimensional Riemannian space V_n, considers a curve given by $x^i = x^i(t)$. The length of the arc is

$$s = \int \phi dt \tag{7.3}$$

where

$$\phi = \sqrt{g_{ij} x'^i y'^j}.$$

This curve is called a geodesic for each i and we can find the Euler's equation

$$\frac{\partial \phi}{\partial x^i} - \frac{d}{dt} \left(\frac{\partial \phi}{\partial x'^i} \right) = 0. \tag{7.4}$$

Now

$$\frac{\partial \phi}{\partial x^i} = \frac{1}{2s \frac{\partial g_{jk}}{\partial x^i}} x'^j x'^k$$

$$\frac{\partial \phi}{\partial x'^i} = \frac{1}{s} g_{ik} x'^k.$$

Therefore,

$$\frac{d}{dt} \left(\frac{\partial \phi}{\partial x'^i} \right) = -\frac{s}{s^2} g_{ik} x'^k + \frac{1}{s} \frac{\partial g_{ik}}{\partial x^j} x'^j x'^k + \frac{1}{s} g_{ik} x'^k.$$

Euler's equation becomes

$$\frac{1}{2s}\frac{\partial g_{jk}}{\partial x^i}x'^j x'^k + \frac{s}{s^2}g_{ik}x'^k - \frac{1}{s}\frac{\partial g_{ik}}{\partial x^j}x'^j x'^k - \frac{1}{s}g_{ik}x''^k = 0. \qquad (7.5)$$

If the length of the arc s as parameter t then

$$s' = 1 \qquad\qquad s'' = 0.$$

The above equation becomes

$$\frac{1}{2}\frac{\partial g_{jk}}{\partial x^i}x'^j x'^k - \frac{\partial g_{ik}}{\partial x^j}x'^j x'^k - g_{ik}x''^k = 0. \qquad (7.6)$$

or

$$g_{ik}x'^k + [jk, i]\, x'^j x^k = 0, \qquad (7.7)$$

where $[jk, i]$ is the christoffel symbol of first kind.

Now transversing with g_{il}, we get

$$\left\{ \begin{array}{c} l \\ jk \end{array} \right\} x'^k x'^j + x''^l = 0, \qquad (7.8)$$

where $\left\{ \begin{array}{c} l \\ jk \end{array} \right\}$ is the Christoffel symbol of second kind. Now we write, equation of geodesic

$$\frac{d^2 x^l}{ds^2} + \left\{ \begin{array}{c} l \\ jk \end{array} \right\} \frac{dx^i}{ds}\frac{dx^k}{ds} \qquad (7.9)$$

Again the arc lenght if p of given as

$$s(t) = \int_{t_o}^{t} |\frac{dp(t)}{dt}| dt = c(t - t_o). \qquad (7.10)$$

It is proportional to the parameter of the geodesic if c=1, the p is normalized. Setting $p(t) = (x_1(t), x_2(t)....., x_n(t))$. By application of the definition, we get

$$\frac{D}{dt}\left(\frac{dp}{dt}\right) = \left(\sum_{ij} \Gamma_{ij}^{k} \frac{dx_i}{dt}\frac{dx_j}{dt} + \frac{d^2x_k}{dt^2}\right) \frac{\partial}{\partial x^k} = 0 \qquad (7.11)$$

which is satisfied, if

$$\frac{d^2x_k}{dt^2} + \sum_{ij} \Gamma_{ij}^{k} \frac{dx_i}{dt}\frac{dx_j}{dt} = 0 \qquad (7.12)$$

for all k. This ordinary, nonlinear second order differential equation is called the Geodesic equation.

7.4 Geodesic Form of the Line Elements

If $\phi(p)$ be a scalar function so that $\nabla\phi = 0, \phi(x) = 0,$ and if u' is the tangent vector along such a geodesic, then we have

$$u' = (dx', 0, 0, 0.......0).$$

The length of vector u' is given by

$$u^2 = g_{ij}u^iu^j$$

$$(dx')^2 = g_{ii}dx^idx^i, g^{ii} = 1.$$

If v' is the tangent vector to the hypersurface, $x' = 0$, then we have

$$v' = (0, dx^2, dx^3, ...dx^n).$$

Also the vectors u' and v' are orthogonal vectors, then $g_{ij}u'v' = 0$

$$g_{ij}u'v' = 0 \quad if \quad (u' = 0 \quad for \quad i = 1, 2, 3.......n)$$

$$g_{ij}v' = 0 \quad if \quad (u' \neq 0)$$

$$g_{ij} = 0 \quad if \quad (v^j \neq 0 \quad for \quad j = 2, 3, ..n)$$

$$g_{ij} = 0 \quad for \quad j = 2, 3, ..n).$$

Now, the differential equation of geodesic is

$$\frac{d^2x^i}{ds^2} + \left\{ \begin{array}{c} i \\ jk \end{array} \right\} \frac{dx^j}{ds}\frac{dx^k}{ds} = 0. \tag{7.13}$$

If t is the unit tangent vector to a geodesic at any point, then

$$t' = 1 \quad and \quad t' = 0 \quad for \quad i \neq 1$$

$$t^i = \frac{dx^i}{ds} = \frac{dx^{i^0}}{dx^i} \qquad \frac{dx^i}{ds} = 1 \quad and \quad \frac{dx^i}{ds} = 0 \quad for \;\; i \neq 1,$$

$$\frac{d^2 x^i}{ds^2} = \frac{d}{ds}\left(\frac{dx^{i^0}}{dx^i}\right) = 0 \quad for \;\; i = 1, 2, \ldots\ldots n.$$

Using Eq.(7.13), we get

$$\begin{bmatrix} i \\ 11 \end{bmatrix} t^j t^k = 0$$

$$\begin{bmatrix} i \\ 11 \end{bmatrix} = 0 \qquad g^{ij}\begin{bmatrix} 11 & j \end{bmatrix} = 0.$$

The line element namely

$$ds^2 = g_{ij}dx^i dx^j$$

$$= g_{i1}dx^i dx^i + g_{jk}dx^j dx^k$$

$$= (dx^i)^2 + g_{jk}dx^j dx^k. \qquad (7.14)$$

Then ds^2 is called the geodesic form of line element.

7.5 Geodesic Coordinate

The components of metric tensors g_{ij} are constant, and then $\frac{\partial g_{ij}}{\partial x^k} = 0$ for every i, j, k. In Riemannian space, there exist a coordinate system with respect to which the christoffel symbol vanishes at given point, and system is called a geodesic coordinate system and the point is called the pole of the given system. suppose a system S whose curvilinear coordinates are u^1, u^2, and a point $P(u^1, u^2)$ on S. If $v^\alpha, \alpha = 1, 2$ are the coordinates of S, then the transformation is given by

$$u^\alpha = u^\alpha(v^1, v^2), \alpha = 1, 2. \tag{7.15}$$

The second derivative formula yields the relation

$$\frac{\partial^2 u^\alpha}{\partial v^\lambda v^\mu} + \left\{ \begin{array}{cc} \alpha \\ \beta \ \gamma \end{array} \right\} \frac{\partial u^\beta}{\partial v^\lambda} \frac{\partial u^\gamma}{\partial u^\mu} = \left\{ \begin{array}{cc} \gamma \\ \lambda \ \mu \end{array} \right\} \frac{\partial u^\alpha}{\partial v^\gamma} \tag{7.16}$$

where $\left\{ \begin{array}{cc} \alpha \\ \beta \ \gamma \end{array} \right\}$ and $\left\{ \begin{array}{cc} \gamma \\ \lambda \ \mu \end{array} \right\}$ are the Christoffel symbols in u coordinates systems and v coordinate system. In Eq.(6.15), $\left\{ \begin{array}{cc} \gamma \\ \lambda \ \mu \end{array} \right\}$ vanishes at P, then for the particular point

$$\frac{\partial^2 u^\alpha}{\partial v^\lambda v^\mu} + \left\{ \begin{array}{cc} \alpha \\ \beta \ \gamma \end{array} \right\} \frac{\partial u^\beta}{\partial v^\lambda} \frac{\partial u^\gamma}{\partial u^\mu} = 0. \tag{7.17}$$

The solution of Eq.(6.17) is transformation of Eq.(6.15) to a coordinate system v^α. The second degree polynomial

$$u^\alpha = u_p^\alpha + v^\alpha - \left\{ \begin{array}{cc} \alpha \\ \gamma \ \mu \end{array} \right\} v^\lambda v^\mu. \tag{7.18}$$

On differentation of (6.18), we get

$$\frac{\partial u^\alpha}{\partial v^\mu} = \delta^\alpha_\mu - \left\{ \begin{matrix} \alpha \\ \lambda \ \mu \end{matrix} \right\} v^\lambda$$

$$\frac{\partial^2 u^\alpha}{\partial v^\lambda v^\mu} = - \left\{ \begin{matrix} \alpha \\ \beta \ \gamma \end{matrix} \right\}_p$$

$$\left(\frac{\partial u^\alpha}{\partial v^\mu} \right) = \delta^\alpha_\mu \quad and \quad \left(\frac{\partial^2 u^\alpha}{\partial v^\lambda v^\mu} \right)_p = - \left\{ \begin{matrix} \alpha \\ \gamma \ \mu \end{matrix} \right\}_p . \qquad (7.19)$$

Using the formula

$$\left(u^\alpha_{\lambda\mu} \right)_p = \left\{ \begin{matrix} \alpha \\ \gamma \ \mu \end{matrix} \right\}_p - \left\{ \begin{matrix} \alpha \\ \beta \ \gamma \end{matrix} \right\} \delta^\alpha_\gamma \qquad (7.20)$$

We see that Eq.(6.19) satisfied Eq.(6.17) at point P and Eq.(618) at P is the new coordinate given by $v^\alpha = 0$ are geodesic coordinate.

7.6 Exercise

(1) Find the geodesics of a sphere of radius u' determined by the equation $x^1 = u^1 sinu^2 cosu^3$, $x^2 = u^1 sinu^2 sinu^3$ and $x^3 = u^1 cosu^2$.

(2) Find the geodesics of a cylindrical coordinate of $x^1 = u^1 cosu^2$, $x^2 = u^1 sinu^2$ and $x^3 = u^3$.

(3) Find the geodesic equation for the metric

$$ds^2 = (dx')^2 + ((x^2)^2 - (x')^2)(dx^2)^2.$$

References

[1] Harold Jeeffreys (1931), Cartesian Tensor, PP(1-66), Combridge University Press (New York).

[2] David C. Kay, Theory and Problem of Tensor Calculus, PP(1-3), McGraw Hill, Washinton, D.C.

[3] Shanti Narayan (1961), Cartesian Tensor, PP(37-51), S. Chand, New Delhi.

[4] Barry Spain (1960), Tensor Calculus, PP(1-55), Dover Publication, Newyork.

[5] Zefer Ahson (2000). Tensor Analysis with Application, Anamaya Publisher, New Delhi.

Index

About the Authors

This book presents basic introduction of tensors with an emphasis on the understanding of the fundamentals. It develops an appreciation of the tensor application in both undergraduate and post graduate students of Physics and Mathematics.

Bipin Singh Koranga is a graduate from Kumaun University, Nainital. He has been with the Theoretical Physics Group, IIT Bombay since 2006 and received the Ph.D. degree in Physics (Neutrino Masses and Mixings) from the Indian Institute of Technology Bombay in 2007. He has been teaching basic courses in Physics and Mathematical Physics at the graduate level for the last 12 years. His research interests include the origin of universe, Physics beyond the standard model, theoretical nuclear Physics, quantum mechanical neutrino oscillation and few topics related to astrology. He has published over 42 scientific papers in various International Journals. His present research interest includes the neutrino mass models and related phenomenology.

Sanjay Kumar Padaliya is presently Head, Department of Mathematics, S.G.R.R. (P.G) College, Dehradun. He received his Ph.D. degree in Mathematics (Fixed Point Theory) from Kumaun University, Nainital. He has been teaching basic courses in

Mathematics at graduate and postgraduate level for the last 20 years. His present research interest includes the Fixed Point Theory and Fuzzy Analysis. He has published over 25 scientific papers in various International Journals of repute and also presented his works at National and International conferences. Dr. Padaliya supervised 05 research scholars for Ph.D. He is also a life member of Indian Mathematical Society, Ramanujan Mathematical Society and International Academy of Physical Sciences.